Capital and the Cosmos

Peter Dickens

Capital and the Cosmos

War, Society and the Quest for Profit

Peter Dickens
Cambridge, Cambridgeshire, UK

ISBN 978-3-031-18500-7   ISBN 978-3-031-18501-4 (eBook)
https://doi.org/10.1007/978-3-031-18501-4

© The Editor(s) (if applicable) and The Author(s), under exclusive licence to Springer Nature Switzerland AG 2022

This work is subject to copyright. All rights are solely and exclusively licensed by the Publisher, whether the whole or part of the material is concerned, specifically the rights of translation, reprinting, reuse of illustrations, recitation, broadcasting, reproduction on microfilms or in any other physical way, and transmission or information storage and retrieval, electronic adaptation, computer software, or by similar or dissimilar methodology now known or hereafter developed.

The use of general descriptive names, registered names, trademarks, service marks, etc. in this publication does not imply, even in the absence of a specific statement, that such names are exempt from the relevant protective laws and regulations and therefore free for general use.

The publisher, the authors, and the editors are safe to assume that the advice and information in this book are believed to be true and accurate at the date of publication. Neither the publisher nor the authors or the editors give a warranty, expressed or implied, with respect to the material contained herein or for any errors or omissions that may have been made. The publisher remains neutral with regard to jurisdictional claims in published maps and institutional affiliations.

This Palgrave Macmillan imprint is published by the registered company Springer Nature Switzerland AG.
The registered company address is: Gewerbestrasse 11, 6330 Cham, Switzerland

Acknowledgements

Many thanks to the following people for their help and support: James Ormrod, Ted Benton, Tristram Hodgkinson, Jo Isaac, the Stronge Family, Matt Taylor, Bryan Turner and Maison Clement @ Derby Street.

Contents

1 Understanding Cosmic Capitalism   1

2 Libertarianism and Cosmic Capitalism   11

3 Narcissism, Fantasy and the Cosmos   25

4 Circuits of Earth, Circuits of Capital   39

5 Working in 'The Silent Sphere of Production'   57

6 Cosmic Capitalism and the Body   73

7 Cosmic Risk Society   95

8 Satellites, War and Capital Accumulation   105

9 Cosmic Capitalism and Space Law   117

10 Future Work: Cosmic Capitalism, Indigenous Peoples and Satellite Television   135

11 Prefigurative Politics: Towards a Cosmic Socialism?   149

Index   155

List of Figures

Fig. 4.1	The primary circuit of capital. (Source: Dickens, 2009)	42
Fig. 4.2	The Primary and secondary circuits combined. (Source: Dickens, 2009)	43
Fig. 4.3	The primary and tertiary circuits Combined. (Source: Dickens, 2009)	44
Fig. 4.4	The primary, secondary and tertiary circuits of capital combined. (Source: Dickens, 2009)	45
Fig. 9.1	Mining the near side of the moon. (Source: Klinger, 2017: 203)	128
Fig. 10.1	Society, nature and the body in indigenous and in capitalist societies	136
Fig. 11.1	NASA's Webb reveals Cosmic Cliffs, glittering landscape of star birth. Jul 12, 2022	152

List of Tables

Table 4.1	Top ten satellite owners	40
Table 9.1	The United Nations Treaty on Principles governing the activities of states in the exploration and use of outer space, including the moon and other celestial bodies	119
Table 10.1	A sample of Imparja TV schedule in July/August 1999	140

1

Understanding Cosmic Capitalism

A major study of society's present relationship with the cosmos is entitled *Scramble for the Skies* (Goswami & Garretson, 2020). The word 'scramble' of course alludes to 'The Scramble for Africa' in which countries competed for the continent's resources during the nineteenth century.

I start with this text because it represents a widely held view about society's relationship with the cosmos, which is quite common. Page One of Goswami and Garretson describes the argument and the tone of their study. Using detailed studies of the countries concerned, they inform their readers that:

> there is a new space race just beginning—though one of a vastly different character. It is not, as before, a race for prestige and honour among nations or a contest between ideologies. It is a rather a race to secure the determinants of economic and military power between states. It a race not just with two superpowers—the United States and the Union of Soviet Socialist Republics (USSR) during the Cold War period—but now joined by the great powers of Asia: China and India. Small financial states and middle powers have also joined in such as the United Arab Emirates (UAE) and Luxembourg. It is a "race" to industrialize the inner solar system through access through access to its billion-fold greater resources. It is a race to write the rules of a new global order. (2020: 2)

Goswami and Garretson go on to describe what they see as the context of this 'race'. This includes an account of the resources of outer space which it terms 'truly vast'. It is referring to 'the mineral wealth of the solar system'. Readers are informed, for example, that 'just one Asteroid, Amun 3554, has platinum, nickel and cobalt worth $20 trillion, a bigger figure than the gross domestic product of Japan, Germany, United Kingdom, and India combined' (2020: 12).

Goswami and Garretson appear stunned by the extent of these resources, and they argue that accessing these resources should proceed as rapidly as possible. Their underlying assumption seems to be that accessing these resources will necessarily benefit the whole of society. (And not, for example, the investors who would govern access and own such resources.)

Goswami and Garretson were working with a version of International Relations Theory. They explain in Appendix A what this means. 'States are theorized to demonstrate interest and investment along a spectrum that begins with elite discourse, becomes instantiated in policy documents, become funded programs leading to prospecting and actual exploitation' (p. 321). Goswami and Garretson go on to give a detailed account of nation-states and their 'resource ambitions'. These states apparently include the United States, China, India, Luxemburg and the United Arab Emirates.

Goswami and Garretson's final chapter asks 'are we observing the beginning of a race or scramble for space resources?' (p. 299). Their application of International Relations theory results in governments and their officials being seen as the main causal powers governing access to the 'the heavens'. They assess 'the salience of perceived possibility of exploiting space resources' (2020: 322). (This means the importance given by governments to space exploration.)

The 'hypothesis' of the authors is that 'as space resources appear to be more exploitable, we will see an increase in competition to exploit those resources by the top tier of great powers' (2020: 322). The authors further distinguish between 'top states' (the United States, China and India) and 'middle powers' benefitting from opportunities that become available.

In short, states are seen by Goswami and Garretson as the causal powers underlying expansion into the cosmos. And 'middle powers' picking up on any opportunities made by 'the top states'. In this way, states are seen by Goswami and Garretson as seen as the causes of interactions, or indeed lack of interactions, between society and outer space. And their interviews of government officials are based on these assumptions. These interviews were based on the earlier-mentioned 'hypothesis' that 'we are observing a shift in the orientation of national space programs and national space policy toward a greater emphasis on the exploitation of space resources' (2020: 328). The 'inferences' behind the research was that 'if we are correct that a scramble is beginning and national space programs are beginning to reorient toward space development and exploitation, then we ought to observe this shift among the great powers who are also spacefaring powers' (2020: op.cit.).

Yet there are surely many difficulties with this perspective. The underlying assumption above is that if a 'scramble' into outer space is taking place, it should be clear by examining the spacefaring political powers and their activities. But 'politics' does not take place in a vacuum. It is surrounded by competing, often irreconcilable, interests, though indeed it is often politics which attempts to make compromises between these interests. In short, 'power' has gone largely missing from this analysis. This is supposedly achieved by 'nation states' and the 'politics' which people can influence through their vote. That is the best capitalism can do in terms of accessing the resources of the cosmos.

But the kind of analysis offered by Goswami and Garretson is surely faulty. The main problem is that it remains stuck at a purely 'political' level of explanation and action, without making the necessary links between politics and capital accumulation. So Goswami and Garretson's line of causation and argument is quite problematic. First, there is no self-evident reason why a scramble for the cosmos should be initiated by governments. In fact, the recent history of space flight demonstrates just the opposite. To an increasing extent, it is the forces of capital, not governments, which is forging society's relations with the cosmos. The old days of NASA firing astronauts up to the Moon are now well gone. And government's job nowadays is to legitimate and support capital and its enterprises. And it is mainly the forces of capital, and not governments, which initiate these adventures into the cosmos. Representatives of capital are frankly not especially interested in extracting the far-distant resources of outer space. It is not at all clear that profits can be made in such a way. But representatives of capital are much more interested in piggy-backing government-led and comparatively small-scale projects, especially those engaged in surveillance and war. So it is the owners capital, not governments, which make military spacecraft and surveillance technologies. And Goswami and Garretson's account offers no understanding of how government policy regarding outer space is made and which interests it might serve.

So in sum, there are two closely related difficulties with the kind of analysis offered by Goswami and Garretson. It limits explanation of a cosmic society to government and government practices. And it thereby ignores the power of capital in the making of space programmes.

Towards an Alternative Understanding: Capital Accumulation and the Military

Society's relations with the cosmos does not consist of spacecraft flying into outer space and harvesting resources. And it is primarily capital, not societies and governments and their spacecraft, which is accessing the cosmos.

So together these relationships completely change our understanding of society's relationship with the cosmos. Capital is using satellites and spacecraft as a means of accumulating profits. And this is being achieved by capital latching on to other interests, including, in particular, those of the military.

The changing role of 'space' in this context is important to remember. In the 1960s, and to a limited extent now, 'space' referred to far distant regions of the cosmos such as the Moon, Mars and beyond. And to a relatively limited extent, 'space' still has this wide meaning. At the time of writing, for example, the U.S. government is planning a return to the Moon for scientific purposes. But this a very exceptional event, one harking back to the 1960s era. 'Space' as now defined by private capital is somewhat more limited. 'Space', furthermore, is increasingly *military* space. The U. S. government in particular uses a presence in outer space to ensure that its citizens remain safe from attack and destruction. Whether such fears are justified is not the point. The real point is that this understanding of 'space' is the understanding adopted by private capital. So now, as later chapters will describe, the military definition and understanding of 'space' is by and large the definition adopted by private capital. This means, for example, that the kinds of hardware used by the U.S. government to keep its people safe are the same hardware as those designed and built by private capital. Capital and government are locked into a real, and potentially very dangerous, embrace. This embrace is described in later chapters of this book.

Capital, Labour, Space and Marx

An alternative to Goswami and Garretson's perspective, one which focuses on the purely 'political' level is to start again, but with a new kind of ontology and a new overarching perspective. As will be clear later, this is based on the perspective offered by Marx. First, he argued that as humanity interacts with and changes external nature it finishes up changes on its own internal nature.

> Labour is first of all a process between man and nature, a process by which man, through his own actions, mediates, regulates and controls the metabolism between himself and nature. He sets in motion the natural forces which belong to his own body, his arms, legs, head and hands, in order to appropriate the materials of nature in a form adapted to his own needs. Through this movement, he acts upon external nature and changes it, and in this way he simultaneously changes his own nature. (*Capital*, 1970: 177)

Developing Marx's concise statement first means getting to grips with the powerful forces of capital, specifically those involved in extending capitalism into the cosmos. But at the same time, it means getting to grips with and understanding human beings' own 'nature'. And as Marx describes in the above quotation, how are the two interacting? And how, as 'man' proceeds into the cosmos, does her or his own internal nature get changed? These matters will be pursued in later chapters.

And these questions raise particularly important questions about our relationships with the cosmos. When I have spoken to people about this book, they assume that the book is about astronauts and 'outer space'. It is marginally about that but it is also about a range of still more important issues. To an increasing extent, as suggested above, capital is being combined with other projects. These include the combination of capital with projects such as the steady investment of capital into satellites bringing television and communications to people back on Earth. So this means that 'outer space' is now a key means by which those with capital and power enhance their power and profits. And to this end, we need to better understand the means by which capital makes its profits in 'outer space' as a whole. As mentioned above, very often these profits are made by spacecraft which only penetrate 'nearby' outer space or by satellites in orbit close to Earth. So a presence in 'space' in this context is a means to an end, that of increasing power relations or enhancing processes of capital accumulation. The following summaries suggest how the above perspective is applied in this study.

Chapter 2 transports the reader to modern-day West Coast America. On the one hand, this area saw the extension of libertarianism, a philosophy of freedom and autonomy with a long history. And this kind of philosophy and politics has been combined and celebrated, particularly in the West Coast of the United States. Libertarian ideals informed the earliest plans to make rockets and extend society into the cosmos.

Chapter 3 describes and explains the narcissistic forces which are another important part of today's relations with the cosmos. This tendency started at the time of the Renaissance when the making of a global economy was accompanied with a new kind of narcissistic human identity. Narcissism now features amongst elites now accessing the cosmos, in large part because it satisfies and inflates their egos. Taken together, libertarianism and narcissism have combined to generate humanity's extension into the cosmos. And when we ask how humanity's internal nature changed as it enters the cosmos, a key part of the answer is the rise of adult infantile narcissism now afflicting human beings as a result of engaging with major capitalist interests and the military.

Chapter 4's focus is twofold. Its main empirical focus is satellites. These are now essential for the transmission of information of many kinds around the world. And they are obviously a keen means of surveillance. The ubiquitous 'satnav' and 'satellite TV' make further links between society and the cosmos. But this chapter also links the so-called satellisation of society to processes of capital accumulation on Earth. Satellite-based communications continue to facilitate the swift rotation of capital essential to capital accumulation.

Chapter 5 describes capital and labour processes involved in making spacecraft. A range of processes is involved. This includes making very large spacecraft designed for a range of civil and military purposes, with latter being increasingly financed by private capital. It also includes making the small landers made by the Jet Propulsion Laboratory and taken in the nose cones of the above large rockets to Mars and elsewhere.

Chapter 6 is concerned with the body in outer space. It discusses the fragile body, the pressures and demands it experiences. It also discusses the problems arising from long distance missions into the cosmos. How does the body cope with these distinctive circumstances? If, as seems likely, there will now be fewer long distance missions into the cosmos, this could mean that the human and its frailties are of less significance. On the other hand, space stations are still circling the Earth, and the body is still exposed to hostile conditions when it proceeds to—or leaves—these stations.

Chapter 7 addresses the question of 'risk' as propounded by Ulrich Beck. He argued that new kinds of risk were opening up as a direct result of modernity. A 'cosmic example' is the increasing amounts of 'space junk' circling the Earth. These are the product of 'modern' and 'successful' inventions such as satellites being used to transfer information and television around the world. But at the same time, they bring various forms of 'risk', this includes satellites colliding with abandoned objects in space and, as a result, causing potentially dangerous accidents within the 'nearby' cosmos.

Chapter 8 addresses the issue of war in outer space. The 'Star Wars' type of scenario—with space 'battleships'—fighting other 'battleships' in the cosmos—is unlikely. But modern wars are not conducted in such a way. Military rockets are typically launched from mobile or aircraft carriers and guided by satellite to their targets. And armies depend on satellite-based information and satellite-guided missiles. Most importantly, this is the key way in which capital now relates to outer space. It is not a matter of society extending indefinitely into the cosmos and accessing its resources. Rather, it is a matter of capital being extended into the cosmos as a result of being integrated into military projects using outer space.

Chapter 9 is about space law. Since the 1960s, UN space law has prevented the private ownership of outer space and the use of the cosmos for military purposes. It remains successful, not least because the pressure to extract rare resources of the Moon, Mars and the asteroids is not—for the time being at least—a priority for capital accumulation.

Chapter 10 suggests future lines of research on satellite television. 'Cosmic capitalism' in this case takes the form of lifestyles and values projected down from satellites to Earth. This chapter covers one particular, but problematic, version of this process. It addresses the problem of indigenous peoples whose cultures and ways of life have often been marginalised by mainstream television. But the chapter finishes on a positive note, showing how indigenous peoples use relatively inexpensive technologies to communicate on a global scale. But this is just one relationship between the public and satellite TV. Of course many others remain to be studied.

The concluding Chap. 11 discusses the possibilities for a cosmic form of socialism, one which allows the cosmos to remain free of capital accumulation. It discusses 'prefigurative' forms of politics in which existing practices on Earth can suggest more progressive ways forward. The new James Webb telescope can be seen as a forerunner of a socialist cosmos, one aimed at enhancing scientific understanding rather than using outer space to create yet more opportunities of warfare and capital accumulation.

References and Further Reading

Davenport, C. (2018). *The Space Barons*. Public Affairs.
Dick, S. (Ed.). (2008). *Remembering the Space Age*. NASA.
Dickens, P. (2009). The Cosmos as Capitalism's Outside. In D. Bell & M. Parker (Eds.), *Space Travel & Culture: From Apollo to Space Tourism*. Blackwell.

Dickens, P. (2010). Alternative Worlds in the Cosmos. In R. Vidal & I. Cornils (Eds.), *Alternative Worlds. Blue-Sky Thinking since 1900*. Peter Lang.
Dickens, P., & Ormrod, J. (2007). *Cosmic Society. Towards a Sociology of the Universe*. Routledge.
Durkheim, E. (1954). *The Elementary Forms of Religious Life*. Allen and Unwin.
Goswami, N., & Garretson, P. (2020). *Scramble for the Skies*. Lexington.
Pathak, P. S. (2022). *The James Webb Space Telescope. The Future of Space Astronomy*. Amazon, Sharma Pashak.
Shammas, V., & Holen, T. (2019). One Giant Leap for Capitalistkind: Private Enterprise in Outer Space. *Palgrave Communications*. https://doi.org/10.1057/s41599-019-0218-9

2

Libertarianism and Cosmic Capitalism

An analogy is often made between the United States opening up the 'West Frontier' in the late nineteenth century and the opening up of outer space in our own times.

The reworking of an imagined past and confronting and conquering a frontier in these ways has been made central to this particular kind of libertarian fantasy. This oft-reinvented tradition, of a toughened individual forged by making a new frontier, has its roots in Frederick Jackson Turner's 1893 *The Significance of the Frontier in American History* (Turner, 1962). He argued that the challenges of the frontier fostered an individualist survivalism, one based on hostility towards centralised power. This in turn leads, Turner believed, to American democracy and the American entrepreneurial spirit. Zubrin and Wagner cite Turner and argued that 'without a frontier from which to breathe life, the spirit that gave rise to the progressive humanistic culture that America has offered to the world for the past several centuries is fading' (1996: 297).

And such visions are key values of today's libertarian right. An imagined past has long been associated with making an essential American national character. Without such a frontier, American culture will, it is seriously argued, stagnate and deteriorate. As we will see, today's space advocates of space travel have quite successfully adopted and adapted this understanding of the frontier, along with imagery, character types and settings evoked to explain and justify the colonisation of the cosmos.

Libertarianism is a philosophy of freedom, one that upholds liberty as its core value. And it is a vision of society which has strongly influenced ideas about a future 'cosmic' form of society. Libertarianism has achieved particular support in the United States. The basic principle is to maximise human autonomy. And this would be achieved by minimising government and thereby releasing human beings' creative capacities.

The Tyranny of the Majority

Yet resistance to over-weaning state power has not always been the dominant concern of libertarianism. John Stuart Mill, the nineteenth century philosopher and a Member of the British Parliament, referred to 'the tyranny of the majority'. He was alluding to the coercive nature of public opinion, one intolerant of any dissidence, eccentricity and difference.

And this wider form of libertarianism remains important today, with the power of 'public opinion' remaining an important target for some of today's libertarians. The preamble to America's Libertarian Party outlines this position as follows:

> As Libertarians, we seek a world of liberty; a world in which all individuals are sovereign over their own lives and no one is forced to sacrifice his or her values for the benefit of others. Our goal is nothing more nor less than a world set free in our lifetime, and it is to this end that we take these stands. (https://www.lp.org/platform/)

By the same token, many libertarians pinpoint government as their chief target, this being seen as the main institution obstructing freedom. Note again, for example, the Statement of Principles of the U.S. Libertarian Party. It starts as follows.

> We, the members of the Libertarian Party, challenge the cult of the omnipotent state and defend the rights of the individual.

This Statement of Principles is fundamental to the American Libertarian Party. It was created specifically to bind the party around some core principles. And note that a very large parliamentary majority is needed by the American party for any such amendment.

But note too that libertarianism has also originated as a form of left-wing politics, this attempting to combine personal liberty with social equality. But the dominant form has developed within the United States from the mid-twentieth century onwards. Here, it has tended to advocate by *laissez-faire* capitalism. And this is important for this story because in the United States, a 'right' form of libertarianism underpinned some of the earliest ideas about space missions.

But today's libertarians also include authoritarian and anti-state socialists. The best examples of these links are anarchists and, more widely, libertarian communists. The kinds of libertarianism adopted depended on who or what the opposition was at the time. These types of libertarianism sought to abolish capitalism and private ownership of the means of production.

So it is clear that libertarian thought can take a range of positions. Furthermore, the same philosopher or activist can adopt a different range of positions according to the particular kind of struggle in which they are engaging. An instructive example in this sense is Karl Marx. A recent study of his work shows how he adapted his 'left' politics according to the state of class forces and class struggles over time. His demand for the dissolution of government, for example, sounds at first like a libertarian or even anarchist position (One in which there is no government). But he saw the removal of government as just a first and transitional stage: that of complete working class liberation. And, for Marx, the next step would of course be the complete removal of the capitalist mode of production organised to produce what was really needed. But Marx managed to combine a clear sense of an end-game (in this case, human liberation) with a more tactical sense of how this end might actually be achieved. So Marx had a clear sense of what was needed to achieve a workers' state. The removal of government was not just a first step towards anarchy. It was a move towards the liberation and the working class taking charge over the whole of their labour processes.

So Marx's political ideals, he argued, would be achieved through transitional phases in which, for example, trade union leaders would not only call for strikes but in due course completely abolish capital and reorganise labour processes within their workplace. Again, self-government as proposed by anarchists would, for Marx, be just a first step towards employees taking over the sphere of production and deciding what it is they want to produce.

The final chapter of this study will further consider the prospects for today's forms of political struggle.

Libertarianism and the Cosmos

So today's libertarianism takes 'right' and 'left' forms. The 'left' version includes socialists, anarchists and others who seek to abolish capitalist social relations and states in favour of common or cooperative forms of ownership. But the 'right' version is the one that has prevailed in the United States from the late 1960s onwards, and this is the version of special interest here because it was this version which influenced the rise of the early plans for outer space missions.

As regards the cosmos and libertarianism in the present era of cosmic capitalism, an important book with a special focus on outer space was R.A. Heinlein's 1966 novel, *The Moon is a Harsh Mistress*. This was a 1966 science fiction novel about a lunar colony's revolt against absentee rule from Earth. The novel illustrates and discusses libertarian ideals. It is still respected for its credible presentation of an imagined future human society on both the Earth and the Moon.

Better-known libertarian theorists today include Christopher Lasch who produced a critical and influential account (1979/1984). He viewed modern American libertarianism as a reaction to the social and political turmoil of the 1960s. These writers in fact encompassed a wide form of libertarianism, one aimed at establishing a more 'authentic' way of life. Lasch (1979) described future forms of libertarian politics as attempting to take charge of a society which has become disconnected from everyday life. The outcome of this disconnection, according to Lasch, has actually been very personalised and unsocial. 'People', he argued, 'have no hope of improving their lives in any of the ways that matter, (they) have convinced themselves that what matters is psychic self-improvement: getting in touch with their feelings'.

But note that Lasch's account and that of Heinlein steers well clear from the sphere of production. In fact, it veered well away from this social and economic forms of politics. And this is surely a major gap in their analysis. Why was—and still is—all this self-improvement and demand for better food and so on taking place? People are not spontaneously demanding healthy food and getting in touch with their inner selves for no reason. They are presumably doing these things, as Lasch implies but does not state, because capitalism does not supply what they need as human beings.

So Lasch's arguments and Heinlein's accounts were acute but at the same time they repeatedly missed the most obvious target: that of economic and political power. This is a serious matter, especially when it comes to politics. Again, demands for better food and decent jobs surely need to start at the proverbial 'factory gate', the sphere of production being the prime site in which people's alienation from other people and from nature starts. What customers get is what corporate capital offers. Yet for all such comments and criticisms, libertarianism remains strong, particularly in parts of the United States.

This means libertarian ideals and politics need to give way to the kind of analysis offered by Marx, one in which alienation (including alienation from decent food and decent work) needs to start with the social relations of industrial capitalism and the labour processes at the workplace. Note that Marx famously referred to 'the silent sphere of production'. This is surely the principle source of unhealthy food and dissatisfying work, but it is a part of social life which attracts little or no interest from libertarians. They have kept 'the silent sphere of production' under wraps.

As mentioned, contemporary libertarianism also takes a flying leap against government rather than capitalism. And this position continues to apply in the United States's 'pro-space' movement and its early form of cosmic capitalism. A right-leaning U.S. version of libertarianism, one which is also deeply engaged with outer space, has recently informed Nelson and Block (2018). Their work is proof that the Californian idealists of the 1960s have definitely not gone away. They are very much still there and they continue to believe that 'market-driven' private initiatives and increased competition between companies will somehow result in the 'colonisation of the planets, moons and asteroids'.

And this process will again supposedly deliver the freedoms (including the freedom to access the cosmos) that corporate capitalism has neglected. Somewhat weirdly but given their politics perhaps predictably, Nelson and Block's book (misleadingly entitled *Space Capitalism*) equates selling things with capitalism. This is again a major error because capitalism is surely not simply about selling things. It is called capitalism because there is something called capital, some of which is owned more by some people than by others. And this is the source of their power. Things of course have been bought and sold in virtually all pre-capitalist civilisations. But obviously did not make Ancient Egypt Ancient Greece or the Roman Empire into 'capitalist' societies. These self-evident facts have somehow escaped from Nelson and Block's analysis.

Again, capitalism is about *making* things as commodities for sale and the social relations formed in such undertakings. Nelson and Block also advocate private and market-driven initiatives and increased competition between companies (where, again, commodities are made) as the way forward for the space industry. Older state-led forms of space exploration, particularly those led by NASA, are said by Nelson and Block to be bureaucratic organisation which have led to 'ossified' thinking about the cosmos and its 'resources'.

Competition between small, more 'agile' companies, Nelson and Block argue without any evidence, are said to bring higher levels of investment into a socialised cosmos with, as a result, high levels of resource extraction. Laissez-faire capitalism as applied to the cosmos would, according to Nelson and Bock, lead to increased resource extraction of raw materials and, as a result, higher levels of satisfaction for everyone. But Nelson and Block do not seem aware of the downsides to their position, specifically the many forms of damage likely to result from such extraction.

In fact, as the present study will suggest a number of times, there is now actually little or no support by capital to start extracting the materials of the cosmos. That is not how capitalism will thrive in the foreseeable future. Such a project would be far too expensive for capitalism to undertake, including as it would, not only mining these resources and bringing them back to Earth. It is a project 'out of this world' and capitalist interests know this very well.

'Capitalism': Some Misplaced Criticisms

It might be expected that 'libertarian' writers such as Nelson and Block would celebrate today's well-known 'self-made' entrepreneurs such as Elon Musk and Jeff Bezos who are now beginning to access the parts of the cosmos nearest to Earth. But this is not the case. Nelson and Block criticise Musk on the grounds that he is not the buccaneering billionaire we might think he is. As they put it,

> Musk's enterprises to reach for the 'stars' and colonise them too cannot be counted as part of the free-market system. Indeed the Musk's project constitute, in the very opposite: economic fascism, government interventionism, crony capitalism. (2018: 190)

So according to this view, Elon Musk, Jeff Bezos and Richard Branson do not represent the kind of 'pure' form of capitalist enterprise that might have appealed to Adam Smith in the late eighteenth century. They and their companies turn out to be deeply dependent on government 'handouts', tax breaks and government projects, all of which are assisting NASA's survival. The latter is a sound point, one indeed made in the following chapters of this book. But Nelson and Block seem to be erasing an earlier period of modern American history when the likes of Musk, Bezos and Branson came into being as entrepreneurs and when today's form of cosmic capitalism got started without massive levels of government support.

Nelson and Block finish up criticising almost everyone, including Musk, Bezos and Branson for their dealings with the U.S. government. And, more predictably, they also take particular offence against United Launch Alliance and other private companies who have benefitted from very substantial state funding. (Chap. 4 of this book also has concerns about United Launch Alliance but for very different reasons.) And,

Nelson and Block suggest, if Musk cannot stand the heat of competition, he and others such as Bezos and Branson should stand aside and let other (unspecified) space entrepreneurs move in with their own plans for new rocketry and new, civilisation saving, missions.

So the notion of 'free enterprise' in space is a dreamy and somewhat unhinged mirage. Nelson and Block again seem strangely unaware of how capital actually operates. Elon Musk first thought about and built a small number of rockets in the 1960s. The U.S. government and the time captured Elon Musk and others, capturing their idealism and recruiting them into projects designed for purposes of warfare and capital accumulation. But these serious forms of cosmic 'usurping' are now being done by big capital and its projects which use investment into spacecraft to undertake warfare, surveillance and capital accumulation and using governments to these ends. This, and not long-distance travelling into the cosmos, is now the way capital is engaging with outer space.

But history and the behaviour of capital is not examined by Nelson and Block. For them, heavy-handed and malicious governments are assuming control over small and defenceless enterprises. But this is turning the world upside down. What is actually happening is capital invested in government projects to further enhance their profits. (These processes and relationship are further spelt out in Chaps. 3, 4 and 5 of the present study.)

By way of a complete contrast with Nelson and Block, Shammas and Holen's (2019) left-leaning account of the space entrepreneurs recognises the topsy-turvy libertarian critique offered by Nelson and Block. But, most importantly, they are coming from a completely opposite political direction. They fully recognise that the original libertarian philosophy informing space travel and exploration has now been thoroughly co-opted and transformed by the considerable forces of capital. Shammas and Holen put the matter deftly.

The frontiersmen of NewSpace tend to think of themselves as libertarians, pioneers beyond the domain of state bureaucracy. But this is now 'space fiction' (2019: 6).

And as Shammas and Holen accurately argue, and as further discussed in later chapters of the present study, 'the entrepreneurial libertarianism of *capitalist kind* is in practice highly dependent on government funding and tax payers' largesse'.

Musk and others are no longer the thrusting independent capitalist enterprises they set out to be. They are now fully incorporated into what is now known as 'the military–industrial–space–complex' (Through, for example, their rockets taking materials to ex-President Trump's newly founded *Space Force*). So somewhat weirdly, the supposed 'autonomy' of the 'private' space initiative actually relies on extensive and expensive government-funded generosity. As Shammas and Holen put it,

> Space libertarianism is libertarian in name only: behind every NewSpace venture looms a thick web of government spending programs, regulatory agencies, public infrastructure, and universities bolstered by research grants from the state and universities bolstered by research. SpaceX would not exist were it not for state-sponsored contracts of satellite launches.

Such are some of the debates and paradoxes surrounding libertarianism and its space programmes today. Despite its origins as a force of resistance and rejection of authority, libertarian has now been systematically combined with the same massive economic and political forces it originally resisted. In this way, it has been thoroughly co-opted into the very power relations and market forces which it originally resisted. And nowadays it is fuelled by a constant blitz of self-publicity by these supposedly 'self-made' men. Meanwhile, as discussed above, Musk's projects have now been combined with government projects. It is a topsy-turvy outcome and highly detached him from the libertarian ideals of 1960s America.

The Cosmos Incorporated into Capitalism

As discussed above and further explained in later chapters, today's capitalism has been made increasingly cosmic. But this kind of cosmic capitalism increasingly depends not only on a physical presence in the cosmos (on, for example surveillance and satellite-based television) but on making strong and profitable links with already-dominant sources of military-cum-economic power. As described in this chapter and as discussed later more detail in later chapters, today's close relations between capital, the military and the U.S. government have been made integral to the extension, strengthening and conglomeration of economic and political power.

Sorensen (2020) describes in great deal the present position, one in which large private corporations are now coming to dominate today's use of the cosmos. And for the forces of capital this is by no means bad news. The GPS system, for example, is officially owned by the U.S. government and it is of course regularly used by many citizens to safely navigate their cars and trucks. Yet Sorensen stresses the corporate ownership of the GPS and allied systems. He sums up the reality of today's corporate conglomeration surrounding space projects as follows.

> The big four—Boeing, Lockheed Martin, Northrop Grumman, Raytheon—hog most design, development and deployment of space-based technology. Global navigation systems (GNSS), upon which people rely in their daily doings, are corporate. The most famous GNSS is the global positioning system, or GPS. Large corporations like Boeing, Honeywell and Lockheed Martin seek GPS provision.

So again, the cosmic visions such as those originally portrayed by Heinlein and Lasch and others have now evaporated and they are fully incorporated into the demands and relations of big capital. Libertarianism still features in American politics, though there must be doubts as to its future as a nationally organised movement. Meanwhile, the libertarian ideals of the original 'new space' movement have been largely reigned in and co-opted into mainstream political alignments and power relations. Elon Musk, for example, is the time of writing being paid by the U.S. government to supply materials to the new and secretive military spacecraft. And he is competing against Bezos's company and three well-established U.S. corporations for the state-funded NASA to make 'the second mission to The Moon'. For sure, the original libertarian ideals connected with outer space are now taking a severe battering.

So again, today's 'libertarianism' is a triumph of form over content. One rocket launch can look much like another but close and critical observers such as Sorensen (2020) and Shammas and Holen (2019) have

2 Libertarianism and Cosmic Capitalism   21

looked behind the smoke, mirrors and spacecraft pyrotechnics to find out what is really taking place. They and others are now detecting and describing the emergence of a new type of cosmic capitalism, one remaining alive and well today but in a form completely different from that described in space fiction. It is now one in which the original ideals of space entrepreneurs are being converted into expensive military rocketry to engage in 'just wars' against imagined enemies. (And making *real* enemies of them in the process.) So again, libertarianism in its earlier forms is now being co-opted and discarded in favour of a government-supported form of cosmic capitalism. But this is not to deny that there are variations within this theme.

Elon Musk has perhaps most obviously detached himself from his libertarian origins. One symptom is his acceptance of NASA (i.e. government) money to carry NASA scientists to the U.S. space station. He is also being paid by NASA (and thanks again from Mr Musk to the U.S. taxpayer who is paying for this enterprise by means of taxation) to take materials to the secretive space station operated by the secretive Space Force spacecraft.

And Elon Musk—and importantly his money—has been recruited by government in the form of NASA to take astronauts 'back to the Moon'. This is the NASA *Artemis* project to design and build a Moon lander financed by NASA and now Mr Musk arriving in late 2024. The claimed purpose is to advance humanity's understanding of the Moon, this rather than using the Moon for military purposes or giving a helping hand to one of the most wealthy individuals on the planet.

And Musk, much like Bezos, still misleadingly portrays himself as a thrusting entrepreneur separate from corporate power while in practice remaining the richest man on Earth and highly dependent on NASA and other government agencies for his 'success'.

Another private company will in due course be made into yet another profit-making opportunity. Astroscale Inc., a private Japanese headquartered company has devised a new spacecraft which will 'declutter outer space'. It is on track to deliver the world's first 'garbage truck' for removing defunct satellites in 2024. Space junk used to be considered a major obstacle for space expeditions. But now this too is being made into yet another sphere of investment and capital accumulation.

Moon Science and Capital Accumulation

A similar kind of corporate intervention is afflicting so-called Moon Science. As mentioned earlier, Jeff Bezos is by some count the richest man in the United States. He founded Amazon, and he still has a substantial shareholding in that company. And his space plans, as discussed shortly, now include giving a helping hand to NASA which is running short of cash for their forthcoming Artemis Moon mission for mainly scientific purposes. Bezos has for some time had a hand in the mainstream space rocket business, this includes making rockets, one of which will be used to make satellites for Earth observation missions and planetary expeditions. And it will also take part in Earth observation missions, planetary expeditions and satellite launches, again all supported by NASA finance.

The Artemis Project: Libertarianism Co-opted

At the time of writing, the high-profile Artemis project, originally invented by NASA aiming to conduct research on physical makeup of The Moon, is running low on research funds and, as a result, it is incorporating private capital into its project.

The list below shows that SpaceX and Blue Origin are two of the five *private* companies tendering to 'mature' (sic) a Moon lander for the forthcoming, NASA funded, Artemis Moon project. It is another example of private capital moving into an originally government outfit. But the welcome is mutual, in the sense that NASA is running low on funds for its scientific programme and urgently needs the incorporation of private partners such as Mr Bezos.

NASA describes and explains its fusion with capital and the forthcoming way. They are there not to make money but to 'help' NASA.

'NASA has selected five U.S. companies to help the agency enable a steady pace of crewed trips to the lunar surface under NASA's *Artemis* program. These companies will make advancements towards sustainable human landing system concepts, conduct risk-reduction activities, and provide feedback on NASA's requirements to cultivate industry capabilities for crewed lunar landing missions'. The five companies are:

SpaceX of Hawthorne, California, $9.4 million.
Blue Origin Federation of Kent, Washington, $25.6 million.
Dynetics, company of Huntsville, Alabama, $40.8 million.
Lockheed Martin of Littleton, Colorado, $35.2 million.
Northrop Grumman of Dulles, Virginia, $34.8 million.

Mr Bezos is upset at the time of writing because NASA has not chosen him to make his Moon lander. In fact he is so upset that he is taking NASA to court because of their unfairness in choosing another private, one that he does not own. Whatever transpires, however, private capital will surely again be the definite winner of this supposed 'government' project. Note that this 'government' project is a made by Lockheed Martin and Boeing.

So libertarianism contains a number or contradictory themes and ideas. Like most of the theories addressed in this study, they need to be seen in relational terms, reacting to and reinforcing each other in sometimes complex ways. But it is clear that the original libertarian ideals, which informed the original U.S. space programme, are now not only being marginalised but made completely absent. But this does not of course spell the end of imaginary phantasies about the prospects of humanity in the cosmos. And as Lothian and Block's study of a future cosmic capitalism shows, a particular kind of libertarianism remains alive, well and increasingly dependent on government funds.

References and Further Reading

Dickens, P., & Ormrod, J. (2007). *Cosmic Society. Towards a Sociology of the Universe* (Chapter Four). Routledge.
Lothian, N. L., & Block, W. E. (2018). *Space Capitalism. How Humans will Colonize Planets, Moons and Asteroids*. Palgrave.
Michels, J., & Lewis, O. (2020). https://blogs.lse.ac.uk/usappblog/2020/12/23/what-happened-the-2020-election-showed-that-libertarians-have-a-long-way-to-go-before-they-can-become-a-national-movement
Nozick, R. (1974). *Anarchy, State, and Utopia*. Blackwell.
Ormrod, J. (2014). *Fantasy and Social Movements*. Palgrave Macmillan.

Screpanti, E. (2007). *Libertarian Communism. Marx, Engels and the Political Economy of Freedom*. Palgrave.

Shammas, V., & Holen, T. (2019). One Giant Leap for capitalistkind: Private Enterprise in Outer Space. *Palgrave Communications*. https://doi.org/10.1057/s41599-019-0218-9

Sorensen, C. (2020). *Understanding the War Industry*. Clarity.

Stone, B. (2021). *Amazon Unbound. Jeff Bezos and the Invention of a Global Empire*. Simon and Schuster.

Turner, F. J. (1962 orig. 1893). *The Significance of the Frontier in American History*. New York, Rinehard and Winston.

3

Narcissism, Fantasy and the Cosmos

This chapter focusses on a particular psychological tendency in contemporary society, one now linked to the idea of space travel and exploration. It is a narcissistic tendency, one in which the self is glorified as part-and-parcel of an imagined expansion of society into the cosmos. When or even whether this cosmic expansion will actually happen is beside the point. These narcissistic tendencies exist as a product of humanity's *imagined* expansion into the cosmos.

Before taking this matter further into our own times, it is useful to return to a much earlier stage of capitalism, one in which humanity was being expanded not over the cosmos but over the globe. A similar kind of self-identity then prevailed amongst dominant classes, and this was a precursor to the kind of personalised identity which was later to develop into a later form of capitalism, one in which the cosmos was being incorporated.

Immanuel Wallerstein traced the rise of the capitalist world economy from the 'long' sixteenth century (c.1450–1640). It was a result of protracted social and economic crises in European feudalism. And capital, what Wallerstein called 'the West', used its advantages to gain control over large parts of the world economy. And from there onwards, it presided over the development of a capitalist society.

And the rise of capitalism was closely connected at the time to cosmopolitanism, a belief that all people are entitled to equal respect and consideration, whatever their citizenship status happened to be. And it was in this context that the earliest forms of capitalism started to develop in some of the core regions of the world.

Burckhardt particularly draws our attention to the cosmopolitanism of that era. And he points us to one particular individual, Leon Battista Alberti (1404–1472), whose 'various gymnastic feats and exercises' attracted much attention in his times. He represented the kind of 'new man' that was to rise with the early development and spread of capitalism over the world.

Burckhardt writes 'in all by which praise is won, Leon Battista Alberti was from his childhood the first. Of his various gymnastic feats and exercises we read with astonishment how, with his feet together, he could spring over a man's head; how in the cathedral, he threw a coin in the air till it was heard to ring against the distant roof; how the wildest horses trembled under him. In three things he desired to appear faultless to others, in walking, in riding, and in speaking' (2012: 94).

So here in Renaissance Florence, a major city at the heart of an early capitalist globalising economy, we find one of the first individualistic selves not only performing superhuman tricks but being widely praised for doing so. And Burckhardt informs us that Roman authors of this time were 'filled and saturated with the conception of fame, and that their subject itself—the universal empire of Rome—stood as a permanent ideal before the minds of Italians'. So in this historical and spatial context, an early form of today's self-centred individualism was featuring in Renaissance Italy, a society at the centre of the earliest form of global capitalism.

Narcissism and Phantasy: Accessing the Distant Cosmos?

Freud was the first to outline and define a narcissistic kind of personality disorder, at a time of course well before humanity started to expand into the cosmos. Infants, he argued, understandably make constant and wholly unreasonable demands on the world, in general, and their parents, in particular, expecting their universe to be oriented around them. This is what Freud called the stage of 'primary narcissism' in which the child is treated, in Freud's phrase, as 'His Majesty the baby' (Freud, 1914: 556). Serious problems result, however, if these narcissistic attitudes persist into later life and the self becomes the chosen love object. This was what Freud called 'secondary narcissism'.

According to Freud, in normal development, people later recognise that they are dependent on significant others. 'Anaclitic' attachments are formed, self-love being displaced onto other people. The family and social life in general also come to impinge on the child's desires, and these result in the child's limitations being internalised. The child becomes gradually aware of the existence of other people, all with their own needs and demands.

But now, well after Freud's death, cosmic capitalism appears to be taking humanity in precisely the opposite direction, back into a form of infantilised adulthood. It is expanding 'His Majesty the Baby', and more particularly, the baby's demand to become a part of the cosmos. And if Freud was correct, this can result in damaging, self-aggrandising tendencies. Meanwhile the harsh fact remains that capital is rapidly investing in 'nearby' outer space. This process remains largely unrecognised and unchallenged.

Capitalist societies such as Britain and the United States encourage desires which are impossible to fulfil. And today's 'idealism', which was once more focussed on altruism (and emancipatory politics) now focuses on the pursuit of *self* expression and the immediate satisfaction of individuals' supposed needs and wants. This process is now being extended as some peoples' needs and wants now imply accessing the cosmos as a means of self-fulfilment. And we are now talking about adults, not children.

Outer Space Narcissism Today

So the expanding kind of subjectivity identified by Freud has continued to thrive, with nowadays fantasies about expanding into the distant 'frontier' of outer space.

As a number of authors have argued, we are now witnessing widespread adult infantile narcissism as a predominant personality type in capitalist societies. As discussed above, Freud (1914) was the first to outline and define this kind of personality disorder. Infants understandably make constant and wholly unreasonable demands on the world, in general, and their parents, in particular, expecting their universe to orient around them. This is the stage of primary narcissism in which the child is treated, again in Freud's phrase, as 'His Majesty the baby' (Freud, 1914: 556). Serious problems result, however, if these attitudes persist into later life as the adult self becomes the chosen love object. (A process known as secondary narcissism.)

This brings us to why this wide scale shift in subjectivity is now happening, a matter which Freud did not foresee. Societies such as Britain, the United States and no doubt elsewhere encourage impossible desires and make reality-testing difficult. Idealism, which was once based on altruism (and emancipatory politics), now focuses on the pursuit of *self* expression and the immediate satisfaction of individuals' supposed needs. Accessing the cosmos is seen as a means of achieving these ends.

Disappointment in this context is usually considered a normal part of psychological development. The id (the unconscious part of the mind from which basic drives emerge) meeting the harsh realities of social relations can therefore be a positive thing. It can harness their excessive demands. But it is increasingly uncommon in late modern capitalism for some groups of people to not to recognise and to attempt over-riding these limitations and tensions.

At this point, we should address pro-space activists into the discussion. These are people, often not very wealthy people, who feel a deep-seated need to access the cosmos. They are amongst those least likely to recognise the inevitable disappointment likely to accompany narcissistic demands.

High levels of consumption are important to recognise here. These fulfil an especially important symbolic role in today's narcissistic culture, again with the consumption of goods providing just the sense of omnipotence that the narcissist craves. And if consumers can continue to make sufficient demands with the aid of their money, they can indeed feel omnipotent and capable of acquiring and achieving almost anything. So in this way, a modern version of Freud's adult infantile narcissism has come to prevail, one fuelled by ready availability of large amounts of money and large amounts of commodities. Taken together, it means that what Freud called 'the reality principle' has still not struck home. The result is a damaging form of self-absorption prevailing, and not only amongst wealthier people.

Narcissism and Cosmic Capitalism

Furthermore, this childlike demand of adults is fuelled by the promise (perhaps in due course the reality) of a cosmic form of capitalism.

The demand for extra goods and/or a flight to the Moon or Mars is what today's narcissistic personality needs. But some kind of rationalisation is needed for those undertaking the trip. Pioneers and proponents of space flight tend to see themselves as adventurers opening up outer space for some kind of 'greater good'. This can include, for example, saving the world from extinction due to climate change. Cosmic adventurers in this context are selflessly saving the world and its people from extinction. This is the plan proposed by Jeff Bezos, for example. So one way in which rationality and fantasy are linked to space exploration is for people to reflect on, and in some way rationalise, capitalism's society's expansion into outer space. It is nothing less than saving civilisation itself.

There is, of course, a well-established counter-hegemonic critique of this frontier mentality, one in which, for example, the destruction caused by the early American Western expansion is highlighted (Launius, 2012). Accessing the Far West, for example, was straightforwardly a way of making a better life. Yet pro-space activists do not apply a critical apparatus to their thinking about the supposed 'necessity' and desirability of capitalist development. 'The frontier', like the difficult notion of 'freedom', remains a transposable myth, particularly in American culture. And it is a 'common sense' that can be very difficult to challenge and dislodge.

Historically, the pro-space movement has had associations with both the political left and right (Kilgore, 2003; Ormrod, 2007). However, coinciding with the rise of private space development projects, this movement has increasingly associated itself with the libertarian right as discussed in the previous chapter.

But again note that pro-space activists by no means see themselves as simply slaves to their narcissistic space fantasies. They actually tend to rationalise their cause and argue in favour of space development, using well-established discourses and tropes about society as a whole and its supposed desperate need for 'rare resources'. These fantasies about fulfilling human 'needs' are largely spurious. (A point developed in later chapters of this study.) But they remain important because they subconsciously inform people's thoughts, actions and politics. They also underlie a process—a mass extension of capitalism into outer space—which is almost certainly not going to take place. The costs involved in making settlements in the cosmos would be, as it were, 'out of this world', but this does not hold back the fantasy about attempting and even achieving such an outcome.

Society, the Cosmos and the Self

Richard Sennett (1979) was one of the first to recognise the above processes and the tensions surrounding contemporary humanity. People, he argued, very much depend on one another and on the social world if they are to survive and make a satisfactory life for themselves and others. But a strong sense of self-absorption now prevails, particularly amongst the more affluent people in the main capitalist societies. And it is one in which the psyche remains relatively immune from the social relations in which adults are inevitably caught up. A result of self-absorption is again a childlike 'rage', one expressing the frustration about not acquiring exactly what the subject really must *have*. Freud's analysis is again useful here.

How do these processes relate to a human presence in the cosmos? James Ormrod, who has studied pro-space activists, argues that these people are not simply engaging in forms of fantasy, escapism and playing out 'forbidden' desires. Rather, fantasy has somewhat more prosaic, but still highly significant, *social* aims. It is 'playing a central role in maintaining the consistency of the subject and the solidarity of the group'.

'The consistency of the subject' here refers to the process outlined by Freud and others. It is one in which pro-space activists and others are subconsciously attempting to recover some of the omnipotence lost as they have become adults. To put this another way, they are engaged in such 'fantastic' and 'omnipotent' thinking which is not simply about getting into outer space. They are attempting to fulfil their childhood fantasies in a process similar to that portrayed by Freud.

So the globalisation of processes set in train during the early Renaissance has continued apace. The private ownership and exploitation of the powers of nature have proceeded very rapidly and often very damagingly. But parallel transformations of internal nature are now taking place. Today's form of individualism still tends to conceive of people as self-directed and preferably owners of substantial levels of cash. And they tend to prioritise independence and uniqueness as their primary cultural values.

Individualistic cultures, especially those in 'The West', tend to conceive of people as self-directed and autonomous, with people prioritising independence and uniqueness. Collectivist cultures, on the other hand, place a premium on people as connected with others and embedded in a broader social context. They tend to emphasise interdependence, family relationships and social conformity. So this is not simply portraying alternative forms of behaviour. As should be clear, and as Craib has suggested, humanity's possible expansion into the cosmos would be a distinctly collective process, one entailing labour processes and very substantial levels of funding. The main point is that adult infantile narcissism, on the other hand, is not only costly but damaging to the adult subject. His or her needs are catered for, but only as a result of accessing high levels of cash from the subjects involved.

As Iain Craib argued, disappointment is a regular and familiar feature of the social life. In practice, nothing is inherently 'good' or 'bad' and whatever might have been planned or hoped-for, experience is usually contradictory. It is usually both 'good' *and* 'bad'.

So a degree of disappointment is inevitable, but this is where human agency and imagination yet again come in. Human beings usually act in the face of disappointment by imagining and doing something else and in this way changing their lives. This is clearly a lesson for everyone and not just for space tourists who may want to access the cosmos. Disappointment is normal to human beings' psychological development. The process of the id (the unconscious part of the mind from which basic drives emerge) meeting with the harsh realities of power is usually good and positive thing. But it is nevertheless common in late modern capitalism for some groups not to recognise the realities of actually existing life. Pro-space activists, those supporting a human presence in the cosmos, are amongst those least likely to recognise these tendencies and the inevitable disappointments likely to accompany their self-centred, narcissistic demands.

Consumption and consumption levels have a particularly important symbolic role in today's narcissistic culture and the 'insatiable personalities' it generates (Dean, 2000). Consuming goods can provide the illusory sense of omnipotence and self that the narcissist craves. They fantasise about their access to the world and its goods, failing to recognise the reality that they are still dependent on other individuals. Yet if they make sufficient demands (particularly with the aid of money), they can appear omnipotent and capable of acquiring and achieving almost anything. For them, Freud's 'reality principle' has still not struck home. And this is damaging in many ways, not only to the individual undertaking such projects but also by other individuals whose rights are over-ridden and unrecognised in the process. Furthermore, it must be added, self-absorption of this kind is also damaging to external nature (in the form of damage to species and the physical environment), as well as to internal nature.

Pro-Space Activism: Accessing the Cosmos to Save the World

With the above kinds of concern in mind, James Ormrod has interviewed pro-space activists campaigning to further explore and develop the universe. Many were fired up by the kind of libertarianism described in the previous chapter. One activist put his reasoning as follows:

> This will be my 8th March Storm, and I come to the March Storm because of love. Love for people, love for freedom and love of opportunity. I have 7 kids, I'd like to leave them the kind of world where they're safe and have the opportunity for prosperity. I love my fellow man, I like to see them happy and prosperous and having a good time. Without the kind of opportunity

that comes from the resources and materials of space I don't see the opportunity being as wide as it could be otherwise. I love freedom. I love the possibility that they can go off into the Universe and make their own kind of lives in their own way without interference by any quantity of other folks who have decided what is best for them.... So I come to Washington to help us unleash the power of the market to change the operating paradigm of space. (2014: 245–6)

As Ormrod points out, pro-space activists (many from the quasi-technical new middle class) are amongst those most affected by late modern narcissism. These people pursue fantasies about exploring and developing space which again manifests the themes stemming from the infant's experience of self during the early stage of primary narcissism. This includes those relating to omnipotence and to unity, with the mother, in particular, and the universe, in general. In these ways, the adult narcissist seeks to regain the experience of primary narcissism and fantasies about conquering and consuming space represent pursuit of this idealised relationship with the universe.

These fantasies have been further encouraged by new developments in space tourism and plans for the private development and settlement of space. And they also achieve a certain legitimacy through the ideology of the libertarian right. Those who have grown up in the 'post-Sputnik' era and were exposed at an early date to science fiction are particularly likely to engage in fantasies or daydreams about travelling in space, owning it, occupying it, consuming it and bringing it under personal control. And Pro-Space advocates often talk about fantasies of bouncing up and down on the Moon or playing golf on it, of mining asteroids or setting up their own colonies.

Clearly, not all of those people growing up in late modern societies come to fantasise about space at an early age. But it is nonetheless a tendency, away in which some dominant sectors of Western society relate to the universe. And it is not only pro-space activists, but many wealthy business people and celebrities who are lining up to take advantage of any new commercial opportunities to explore space as tourists and of other ways of symbolically consuming the universe. The promise of power over the whole universe is in this way, therefore, the latest stage in the escalation of the narcissistic personality.

A new kind of 'universal man' is in the making. Space travel and possible occupation of other planets further inflate people's sense of omnipotence. This despite the fact (or even because of the fact) that it is not actually taking place. Indeed, as demonstrated later in this study, precisely the opposite process is taking place. Capital is now above all being expanded into the 'nearby' cosmos.

Fromm (1976) examined how in Western societies people experience the world (or indeed the universe) through the 'having' mode. Extending Fromm's insight to the present day, we are now witnessing a time when individuals (wealthy individuals in particular) cannot simply appreciate the things such as the cosmos around them. Rather they must own them, mine them and use them to make a swift profit. Cosmic capitalism caters for these sentiments despite the fact that these visionary phantasists are very unlikely to see their desires fulfilled.

In 2004, Mean and Wilsdon made the causal connection between the disenchanted universe viewed only as objects and a particular kind of consumerism.

> The underlying anxiety and disorientation that pervade modern societies in the face of a meaningless cosmos create both a collective psychic numbness and a desperate spiritual hunger, leading to an addictive, insatiable craving for ever more material goods. (See Dickens & Ormrod, 2007b: 75)

For the narcissistic pro-space activist, this sentiment means they feel a desperate need not just to look at the Moon but to have immediate sensuous contact with it, and thereby somehow bring it closer to their control. James Ormrod talked with these people, and it received the following explanation of their projects.

> Some people will look up at the full Moon and they'll think about the beauty of it and the romance and history and whatever. I'll think of some of those too but the primary thing on my mind is gee I wonder what it looks like up there in that particular area, gee I'd love to see that myself. I don't want to look at it up there, I want to walk on it. (25-year-old engi-

neering graduate interviewed by James Ormrod at a ProSpace March Storm 2004)

A 46-year-old space scientist interviewed at ProSpace March Storm 2004 put the matter of access to the cosmos as follows:

> It really presents a different perspective on your life when you can think that you can actually throw yourself into another activity and transform it, and when we have a day when we look out in the sky and we see lights on the Moon, something like that or you think that I know a friend who's on the other side of the Sun right now. You know, it just changes the nature of looking at the sky too.

Note that this kind of demand for access to and ownership of the cosmos can be *partly* satisfied by individuals now thinking about purchasing a star. (The International Star Registry is currently offering stars for £14.99 each.) Here is another way in which experience of the cosmos is fundamentally changing. It is now possible to actually buy and own parts of it and without actually going there.

Cosmic Narcissism as Self-ownership

A widespread cosmic narcissism of this kind might appear to have an almost spiritual nature, but the cosmic spirituality we are witnessing here is not about becoming immortal and connected to the heavens. Rather, it is in the end, a spirituality taking the form of *self*-worship. It is further aggrandising the atomised, self-seeking, twenty-first century individual by connecting her or him to the heavens. For cosmic narcissists, the universe is again very much experienced as an object: something to be conquered, controlled and consumed as a reflection of the powers of the self. This vision is inherited from the assumptions made by Francis Bacon about the relationship between man and nature on Earth. The cosmos, like nature more generally, is there to be conquered and its resources extracted for human use. However, narcissistic relationships with external nature are intrinsically unsatisfying. In practice, the objectification of nature and the cosmos does not actually empower the self. Rather, it enslaves it.

Pro-spacers' lack of the reality principle presents itself in a number of quite disturbing ways. Many activists had wanted to be astronauts but had been turned down. Yet the first barrier of not meeting the requirements of a governmental programme has not dampened their enthusiasm. Within the U.S. space programmes, only a small elite got to fulfil these dreams. Now, private industry (particularly the companies owned by Elon Musk and Jeff Bezos) are planning for many more people to have an opportunity to realise their dreams and childlike phantasies.

Later chapters will be developing some of the themes in the context theories about 'civil society', the realm of social life outside the sphere of production. In these cases, fantasies about occupying and living in the cosmos are a product of not only 'fantastical' thinking but in the present case by a very few immensely wealthy people describing the form and extent of real, actual fantasies.

The harsh fact is that travel into the cosmos really is a fantasy. A few people may get to the Moon, Mars or an asteroid for a short while in the distant future. But this is a fairly trivial matter. The harsh fact is that social and political life will remain firmly on Earth for any foreseeable future. And, if the argument of this book is correct, it is primarily capital—combining with the government and especially the military—which is determining how the cosmos, and particularly the 'nearby' parts of the cosmos, will be used. If we find this reality 'disappointing', we would be better off opposing capital and the power relations involved rather than fantasising about cosmic alternatives.

References and Further Reading

Burckhardt, J. (2012). *The Civilization of the Renaissance in Italy*. Renaissance Classics.
Chodorow, N. J. (1999). *The Power of Feelings. Personal Meaning in Psychoanalysis, Gender and Culture*. Yale University Press.
Craib, I. (1994). *The Importance of Disappointment*. Routledge.

Dickens, P. (2001). Changing our environment, changing ourselves. *Journal of Critical Realism, 4*(2), 9–18.
Dickens, P. (2003). *Society and Nature. Changing Our Environment, Changing Ourselves.* Polity.
Dickens, P., & Ormrod, J. (2007a). 'Outer Space and Internal Nature'. Towards a Sociology of the Universe. *Sociology, 41*(4), 609–626.
Dickens, P., & Ormrod, J. (2007b). *Cosmic Society. Towards a Sociology of the Universe.* Routledge.
Freud, S. (1914). 'On Narcissism: an Introduction' in *The Complete Psychological Works of Sigmund Freud*, vol 14 (pp. 217–235). London, Hogarth.
Freud, S. (1986). *New Introductory Lectures on Psychoanalysis.* Pelican.
Higgs, J. (2021). *William Blake vs The World.* Weidenfeld.
Jackson, R. (1981). *Fantasy. The Literature of Subversion.* Routledge.
Kilgore, D. W. D. (2003). *Astrofuturism. Science, Race and Visions of Utopia in Space.* University of Pennsylvania.
Klein, M. (1975). Love, Guilt and Reparation. In R. Money-Kyrle (Ed.), *Love, Guilt and Reparation and Other Works.* Delacorte.
Lasch, C. (1979). *The Culture of Narcissism.* Norton.
Launius, R. D. (2012). Compelling Rationales for Spaceflight? History and the Search for Relevance. In S. J. Dick & R. D. Launius, *Critical Issues in the History of Spaceflight.* N.A.S.A. Washington DC.
Ormrod, J. S. (2007). Pro-Space Activism and Narcissistic Phantasy. *Psychoanalysis, Culture and Society, 12*, 260–278.
Ormrod, J. (2014). *Fantasy and Social Movements.* Palgrave Macmillan.
Ormrod, J. (2016). Space Activism: A Psychosocial Perspective (Chap. 14). In P. Dickens & J. Ormrod (Eds.), *The Palgrave Handbook of Society, Culture and Outer Space.* Palgrave.

4

Circuits of Earth, Circuits of Capital

This chapter outlines a framework for understanding the spread of capitalism into the cosmos, this including 'nearby' parts of outer space. It gives special attention to what is sometimes called 'the satellisation of society'.

Satellites are now a central feature of everyday life. They are, for example, a key feature of mobile phones. But it is important to grasp the many other connections and processes involved.

The following list gives a rough idea as to who or what is in charge of satellites and their use. It shows the 'top twelve' owners of orbiting satellites in 2022, and it shows that satellites are very closely linked to organisations with major social, economic and military power. These satellites and their links are therefore particularly instructive if we are to understand the importance of the cosmos for contemporary capitalist society.

Table 4.1 The top twelve satellite owners

Table 4.1 Top ten satellite owners

1. Space X (Elon Musk)
2. Planet Labs Inc. (private U.S. Earth imaging)
3. Chinese Ministry of Defence
4. Spire Global Inc. (maritime/aviation data)
5. Swan Technologies (web design and marketing)
6. The United States Air Force
7. Iridium Communications Inc. (voice/data coms)
8. National Reconnaissance Office. (U.S. surveillance)
9. National Aeronautics and Space Administration (NASA)
10. European Space Agency (ESA)

Source: DEWESoft (2022)

Satellites: Their Contradictory Purposes

Table 4.1 gives a rough idea about the use of satellites. But there are still more important processes taking place here, and these are the main point of this chapter. Satellite production, like spacecraft production, is closely linked to surveillance and warfare. And governments, especially the U.S. government, deeply depend on private capital to prepare for, and if necessary conduct, wars. But whether or not military hardware is actually used for these purposes is not even the main point. Capital, as will be discussed throughout this study, is looking to make a profit. And extensive government spending on military and surveillance hardware (again made by private capitalist enterprises) is a central means to this end. Capital, in other words, is deeply engaged in war and surveillance, and this is why an important point to which this study will return.

Satellites are used, for example, to guide military cruise missiles capable of delivering high-explosive bombs over long distances. And satellites also feature in international politics by monitoring military movements and targeting suspected 'enemies' for their destruction. (At the time of writing, the Russian Invasion of Ukraine is an example. It is using satellite for purposes of observation and attacks on military and non-military assets.) And note that private capital is the main beneficiary of these 'military' enterprises. Capital is the chief beneficiary, this because capital makes, and when necessary replaces, military weaponry.

With the above relations and processes in mind, this chapter examines some of the many, and sometimes less obvious, ways in which satellites are linked to capitalist society. To this end, it adopts and extends the work of the geographer, David Harvey. He explains the expansion of capital on a global scale and the relations and processes involved. Harvey's analysis is extended here as a means of understanding the spread of capitalism on a cosmic scale.

Capitalism, the Labour Processes and Satellites

For many years, satellites have been routinely bringing access and have shared all kinds of information to humanity on Earth. The omnipresent and now taken-for-granted 'satnav' car guidance system, for example, critically depends on satellite communications. And satellite-based mobile phones, once considered a luxury item, are almost universally owned and used. These are all examples not of capital accumulation per se but of capital 'piggybacking' on Earthly needs to—including military 'needs'—to make profits. And in this case, it is the need for people to communicate with one another, sometimes over vast distances.

Note the implications of Elon Musk's new Starlink project in the above table. It is based on 42,000 satellites forming a 'megaconstellation' and the idea is that it will bring low-cost internet services for businesses and citizens. Whether or not Musk's ambitious project for the whole of the world remains to be seen. But it is another example, not of humanising the cosmos per se but using a presence (in this case, a nearby presence) in the cosmos as a means of enhancing levels of capital accumulation.

Cosmic 'Satellisation' and Circuits of Capital

Satellites are extensively used to exercise social and political power. Their presence in the cosmos may be less exotic than spacecraft but for sure they are a key means by which capitalism is being reproduced. They are used, for example, to track employees and to enhance levels of productivity. This even when employees are physically well-distanced from their workplaces. Employees are constantly under surveillance, with the underlying aim of making them work more productively. Mobile phones offer a way of keeping 'tab' on employees and 'satellisation', the presence of satellites in the nearby cosmos, is an important element of this process.

To understand the key underlying processes generating the expansion of capitalist society into the cosmos, we need to turn back to David Harvey's account of how and why capitalism is spreading around the world. Again, his perspective on capital accumulation on a global scale can be extended to understand how and why capitalism is being spread via communication satellites in the 'nearby' cosmos.

The 'primary circuit' of capital in Harvey's scheme is represented by Fig. 4.1, this of course being the heart of capitalist society. Capitalism, whether on Earth or extended into the Cosmos in the form of spacecraft and armaments, starts here.

Figure 4.1 represents the centre of capitalism, one in which capital and labour are invested in 'the labour process' (or in Harvey's words 'work process') to produce the surplus value left over when the capitalist has taken a profit. ('Surplus value' is the difference between the amount raised through by the sale of a product and the amount it cost the owner of a factory to produce and sell the product.) So in this way, labour processes are at the core of capitalism. Capital is ploughed into labour processes (or the 'work process') to produce the commodities people can afford to buy (Left side of Fig. 4.1).

Labour processes are sometimes forgotten, even by the important critical account of cosmic capitalism recently introduced by Shammas and Holen (2019). Labour processes, like indeed satellites, are out of sight and consequently often out of mind. Capital's 'fixes' in space on Earth are not just made for their own sake. Rather, they are absolutely central to the labour processes which generate profits. 'Fixes' of capital per se, and made without labour processes, would result in rapid and calamitous bankruptcy for factory owners and capital investors alike.

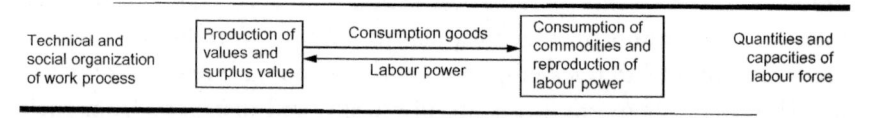

Technical and social organization of work process	Production of values and surplus value	Consumption goods ⟶ ⟵ Labour power	Consumption of commodities and reproduction of labour power	Quantities and capacities of labour force

Fig. 4.1 The primary circuit of capital. (Source: Dickens, 2009)

Circuits of Capital and Cosmic Fixes

Labour processes are currently based on the extraction of rare materials on Earth. But here lies a critical question for our interest in the cosmos. 'What materials are in outer space and who actually owns outer space?' Outer space has been conceived as *terra nullius* (territory as yet unclaimed but claimable) or (*res humanitas*), owned by everyone. But it is noticeable that no such claim of ownership has yet been made. This again undermines the popular notion that capitalism is due to start spreading into far distant outer space anytime soon. It may do so in the distant future but for now capital is more than happy to be closely integrated into the 'nearby' cosmos which is host to a range of agencies (governmental as well as 'private') using satellites in the 'nearby' parts the cosmos. Satellites in 'nearby' outer space are now routinely used as a means of capital accumulation and surveillance. It is not a particularly exotic or inspirational picture of humanity's relations with the cosmos but that is how it is.

So capital, for the time being at least, is more than happy to continue investing to use and invest in these purposes, not least because governments are guaranteeing a stream of income derived from surveillance and war. But at the same time, 'work processes' are technically and socially organised in such a way as to produce profits stemming from 'normal' forms of consumption (right-hand side of Fig. 4.2) and for the reproduction of 'labour power' or the potential to work. In short, governments are there to facilitate the making of profits. And this can be achieved in a number of ways including preparation for wars and real wars.

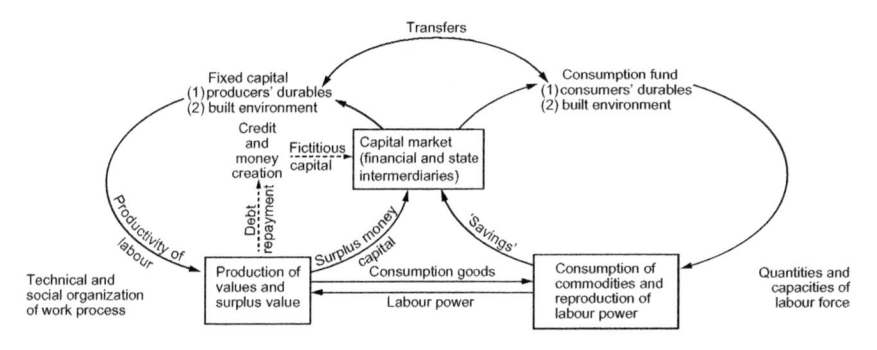

Fig. 4.2 The Primary and secondary circuits combined. (Source: Dickens, 2009)

Figure 4.2 builds up the wider picture. It shows the main 'secondary circuits' linked to the primary circuit of capital portrayed by Fig. 4.1. Surpluses of capital stemming from production process here are, via the capital market, used to finance the making of the built environment and 'produce durables' (equipment and machinery of all kinds). On the right hand side of Fig. 4.2, we see household 'savings' being invested back into the capital market where financial experts invest these 'savings' in ways which, hopefully for investors, secure a satisfactory return.

Surveillance of labour processes in the service of capital accumulation has taken place at least since the early days of the industrial revolution. But such monitoring now takes new forms and again often relying on satellites. One type of satellite-based monitoring, for example, entails employees wearing small 'tagged' transmitters which can be monitored by satellite (Austermuehle, 2016). In this way, employees can be encouraged to work harder and more productively.

High-speed computing also nowadays uses specialised hardware and software connected to Ground Positioning Systems (GPS) to monitor the location of employees. And such screening is also often conducted via satellite, this allowing measurement of productivity. In such ways, satellites have been used by managers to track labour and productivity for at least two decades. And cosmic capitalism has been a means to this end.

Turning now to Fig. 4.3, further surpluses here are being drawn by taxes from households and companies to support their 'state functions'. These include investments in education and welfare which will hopefully enhance a nation's well-being and productivity at the workplace. Also included here are education and welfare assets intended to create a

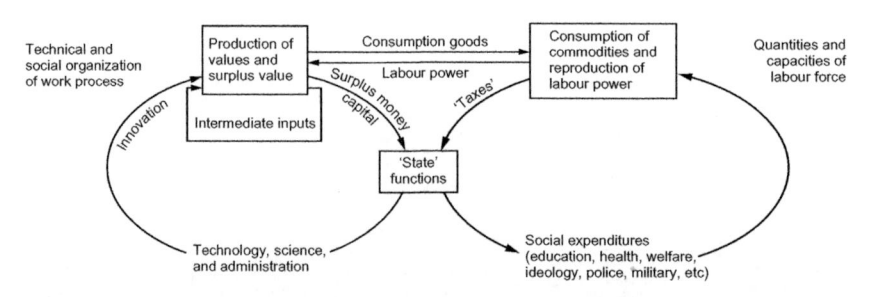

Fig. 4.3 The primary and tertiary circuits Combined. (Source: Dickens, 2009)

healthy—and hopefully 'productive'—labour force. Meanwhile direct investment is made into military assets to 'keep the peace' while enriching private investors in this sector. This is an important issue to which this study will return.

On the left side, these surpluses are being reinvested in technology, science and government administration. On the right side, taxes are being extracted from consumers by means of government taxes. And they are recycled into a range of government expenditures including welfare provision as outlined earlier. These investments trigger still more chains of commodity production as illustrated in Fig. 4.1. Again, the labour processes involved may be monitored via satellite or some other form of surveillance to enhance productivity at the workplace.

The whole process is summed up in Fig. 4.4. showing the primary, secondary and tertiary circuits of capital combined, with the production and consumption of goods and services at the heart of the whole process.

Figure 4.4 also represents endless processes of creation and destruction, with 'fixes', labour-processes and commodities being made and, over time, replaced by new 'fixes' creating new labour processes and

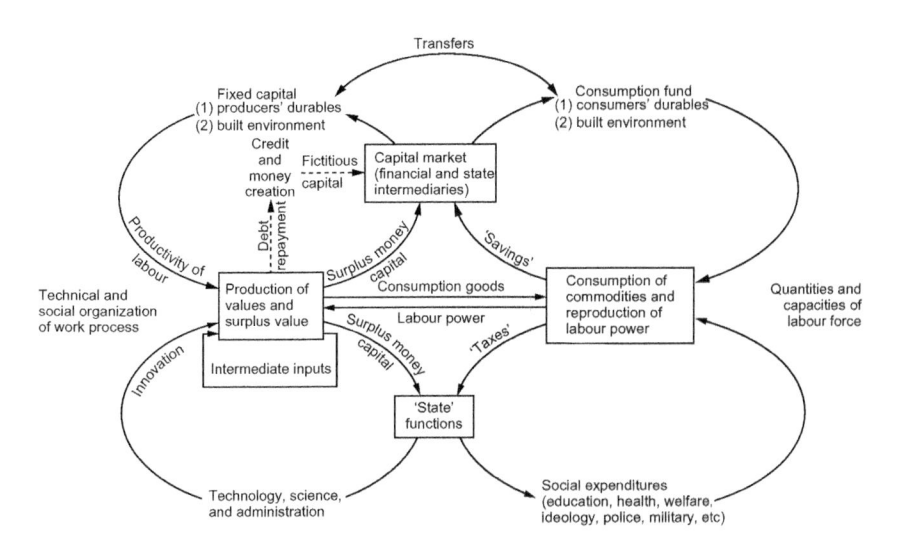

Fig. 4.4 The primary, secondary and tertiary circuits of capital combined. (Source: Dickens, 2009)

products all with, of course, the aim of securing profits (leaving behind, we should add, the large and increasingly dangerous amounts of 'space junk' circling Earth). The overall picture is one of a rapidly expanding capitalism, with surpluses drawn off and turned in profits for company owners and shareholders. Other surpluses will be recycled into yet more circuits of capital and labour processes. Satellite communications are central to this process.

Circuits of Capital, Labour Processes and Surveillance

How in more detail do the circuits of capital by satellites relate to people on Earth? I now turn to the labour processes at the heart of the above circuits of capital. These are key to understanding the accumulation of capital and, as part of this process, capitalism's extension into the cosmos with satellites.

Employees are typically placed under some form of surveillance, this again with the intention of increasing productivity and profitability. Surveillance can take various forms, again many relying on satellite communications. As Blakemore's extensive research showed, satellites are widely used for employee surveillance. Blakemore (2011). More recently, Wills (2016) has described in great deal the satellite surveillance systems used to monitor military and non-military practices.

The primary circuit as laid out earlier in this chapter is the key and central process of capitalism so it comes as no surprise that employees in this sphere are routinely monitored (Lazzarato, 2020). Surveillance takes varied forms, again many relying on satellites. One type of satellite-based monitoring entails workers wearing a small 'tagged' transmitter which enables employers to keep tab on their employees. In the early days of satellite surveillance, employees were regularly monitored in such ways, the underlying aim of course being to induce them to labour productively.

The impacts of such surveillance are quite subtle. Employees are likely to be told they *may* be under surveillance. And this acts as way of making employees into self-monitoring and hence productive individuals. There are definite hints throughout this literature of the kind of surveillance described by Michel Foucault (1977) in relation to prisons. People 'behave well' because they *think they may be* under surveillance.

But there is another even more subtle and more disturbing form of control over labour processes. Satellite-based communications systems are also now used to track 'immaterial labour' (Lazzarato, 2020). This refers to recent developments in which the fundamental aspects of human social life—including even people's use of language and discourse—are recognised, incorporated or 'subsumed' under capital and made into one of its most important assets. These 'assets' are not simply language. They are social relations and behaviours which require capturing, co-opting and as far as possible using towards the overall processes of capital accumulation.

Today's form of labour process (particularly those in the 'services' sector which often involve direct contact between employees and consumers) places a special premium of people's communication abilities. And this brings us back to satellites because, as mentioned earlier, many employees are issued with mobile phones and other means of communication. These include 'personal digital assistants', which enable regular contact between managers and employees. Such technologies have a dual use, particularly in the 'services' sector. They are obviously important as a means of delivering and taking instructions. But they are also important as a means by which managers can monitor employees' ability to communicate and assist customers.

Communication and tracking is especially important to employers. High-speed GPS (Ground Positioning System) satellite-based communications have for some time been important for surveillance by managers. And GPS is an important means to these ends. Similar processes are evident in the workplace where 'Just-in-Time' planning entails manufacturers and subcontractors remaining in close and regular contact. GPS communications also help company managers to keep a close eye on stock levels and reduce warehousing costs.

Managing by Satellite: Some Contradictions Arising

In all the above ways, management of labour via satellites and the Ground Positioning System remain central to contemporary labour processes and forms of capital accumulation. But, importantly, they can sometimes be counterproductive. 'Reduction of stocks', as Brin (2009) puts it, 'can be adhered to as an almost religious mantra, even to the extent of creating lost profits'. The result can easily be what Ritzer (2013) called 'the irrationality of rationality'. He is referring here to attempts to increase efficiency, predictability and control over labour processes which can clutter up labour processes and perhaps slow down the conversion of employees labour into commodities.

For example, a two-year delay in the production of the Boeing Corporation's new 787 Dreamliner aircraft was largely the result of an over-rigid dependence on Just-in-Time (JIT) production, with a system based on over-optimistic assumptions about knife-edge delivery times. But deliveries were late and the delicately balanced just-in-time production process failed, thus threatening productivity and profits. JIT has also been adopted in the public sector, including NASA's space programme. But in this case, it was severely undermined, in this case by an over rigid adoption of this strategy. JIT, managed by satellite monitoring, can have perverse results.

So, while satellite-based communications serve to increase the power of owners and managers, satellite-based communications systems are not necessarily unproblematic and risk-free. Seemingly 'sophisticated', 'rational' strategies of monitoring labour processes designed to increase productivity can actually bring about increased levels of risk confusion and even failure.

Roberts and Armitage (2006), whose work draws on that of Virilio, argues that contemporary society is characterised by 'hypermodernism', a phenomenon in which information and communication technologies (ICTs) are capable of operating at extraordinarily high speeds. The work of Virilio and his followers is a stimulating commentary on the speeding up of every day life but it is not helpful for analysis to focus on 'speed' alone. Again, speed, or 'acceleration' is not necessarily an end in itself. It is best seen as a means of increasing productivity and capital accumulation. Rapid rotation of capital—into and quickly out of—labour processes is obviously important to managers. But for sure, this rotation may not work in the way intended and it may well of course be resisted by those under overintensive surveillance. But the main point here is that these processes, successful or otherwise, critically depend on the use of satellites circling Earth.

Satellites, Lifestyles and 'New' Forms of Consumption

Returning now to the primary circuit of capital, commodities are next sold on the market. To stop the circuit of capital grinding to a halt, consumers must be persuaded that they need to abandon their old purchases and adopt 'modern' lifestyles and 'modern' forms of consumption. Lifestyles and forms of consumption are therefore promoted to persuade people that they actually *need* these commodities. But where does this promotion and persuasion come from? And how is the cosmos implicated? Capital's access to satellites is an important part of the explanation.

Much of this 'lifestyle' promotion is, particularly in developing countries, conducted via 'Direct to Home' (DTH) TV broadcasting from satellites in geostationary orbit. An example is Direct TV based in Latin America. Three satellites in geostationary orbit 22,000 miles above Earth are sufficient to broadcast worldwide. In the 'advanced' societies, broadcasting is conducted by a combination of satellite and cable networks operating at a global scale. As a result, people in so-called developing countries are rapidly opened up as places to promote and sell 'Western' and consumerist ways of life and values.

In sum, the speed made available by satellite communications is again used towards a definite end, again that of rapid capital accumulation, with profits made in one round of production being rapidly transferred into other circuits and other rounds of accumulation and so on ad infinitum. In such ways, cultures, especially dominant cultures in the United States but elsewhere, are promoted and to a degree accepted by TV 'consumers'.

These processes of investment, realisation of profits and reinvestment must be conducted by the capitalist as rapidly as possible. As Harvey puts it, 'the faster the capital launched into circulation can be recuperated, the greater the profit will be'. And companies are increasingly obliged to follow the latest 'speedy', often satellite-based, technologies if they are to remain profitable. In these ways, the cycles of production, distribution and consumption is under constant pressure to accelerate. Innovations dedicated to the removal of spatial barriers and to enhance speed have been immensely significant in the history of capitalism. Railroads, the telegraph, the automobile, radio, telephone, jet aircraft, television and now satellite-based telecommunications are used to reduce the importance of time and distance to everyday life. Again, this speeding up process is not an end in itself. The point is that 'speedy' innovations are used to improve capital's turnover times, with new technologies and products being invented, marketed and sold by capitalist companies. Speedy rotation of capital is a key way of realising profits from investments.

But again, the 'relative surplus value' produced by such rapid technological innovation is not guaranteed for ever. Technological innovation may result in an increase in market share and profitability for a while, but in due course, other entrepreneurs will catch up with the same or even more profitable innovations and levels of speed. They must do so to stay in business.

Satellite communications, with their capacity for transmitting information over long distances at the speed of light, remain a key part of today's surveillance technologies aimed at the consumers under continuing pressure to purchase new goods. And television has a large role to play in encouraging people to buy more commodities. But later in Chap. 9, we will find groups of peoples who have largely opted out of persuasion-by-television and who have found more satisfying ways of using their time.

Satellites and Capital Flows

The primary circuit described as the start of this chapter is, as we have seen, linked to what Harvey calls the 'secondary circuit of capital' (2003). 'Surplus money capital' is being circulated via the capital market into 'fixed capital'. This includes the machinery needed for production processes and infrastructures such as buildings, roads and satellite communications systems. The secondary circuit links to a vast range of further primary circuits where profits are realised. Satellite communications, as outlined above, are nowadays central to capital accumulation and the promotion of consumer spending.

Finance Capital and the Cosmos

The financial sector is especially important to note here. The role of capital markets is to switch the surpluses made in the sphere of production and the savings made by households. And now rapid switching of finances is even resulting in stockbrokers losing their jobs. This is partly because they cannot invest capital quickly enough to take advantage of new opportunities that rapidly arise. As Duhigg puts it,

> powerful computers enable high-frequency traders to transmit millions of orders at lightning speed and, their detractors contend, reap billions at everyone else's expense. Powerful algorithms "algos" in industry parlance—execute millions of orders a second and can scan dozens of public and private marketplaces simultaneously. They can spot trends before other investors can blink, changing orders and strategies within milliseconds.

And the 'high frequency trading' conducted in this sector is highly dependent on satellite communications. For example, traders use GPS technology for accurate time signals, these being an important factor in trading strategies and making deals. Electronic Funds Transfers, in which money is moved very rapidly between one bank and another, are routinely made by satellite as well as by fibre optics (Warf, 2013).

So speed in the flourishing financial sector, with its rapidly fluctuating money values and its demand to circulate capital as rapidly as possible, is of the essence. And for this reason, electronic communications via satellites are an integral part of this process. The rapidity at which financial data is transmitted can, in the words of one commentator, 'make millions of dollars worth of difference to high-frequency traders'. After all, the ultrafast rotation of capital is a prime means by which financiers enhance their own profits and achieve what Michl called their 'insatiable quest for high incomes'. Finally, note that high-speed computer trading tends to strengthen the already-strong social position of what Van der Pijl calls 'the new empire of high finance'.

In these ways, the power of the financial sector, a power necessitating swift satellite communications, greatly enhanced investments and disinvestments into and out of stocks and shares. In these ways, nearby 'outer space' is fully, but almost invisibly, implicated in the process of capital accumulation.

But again, none of this is to deny there can again be major glitches. The new satellite technologies can also generate their own kinds of crisis. Some financial centres without adequate security back-up to their computers have, for example, been temporarily shut down as a result of satellite failures. And in September 2006, India's National Stock Exchange was even closed by 'sun outages', the overwhelming of satellite signals by high levels of solar radiation.

Further unanticipated crises affecting the financial sector and the economy as a whole can be created by ultrafast forms of investment via satellite. Virilio reminds us that each new and faster form of communication made by using satellite communications can contain the potential for a new form of what he calls an 'integral accident' (Crosthwaite, 2011). New, and very speedy, 'integral accidents' have indeed been generated in the financial sector. One of the best-known examples was the so-called flash crash in 2010 when, due to a glitch in satellite connections, share prices of the some of the world's largest companies fell by up to 99% in under half an hour.

So the trading strategies of high frequency traders are highly dependent on GPS-led technology. But the fact that such trades can now be conducted solely by computers and satellites, with little or no human intervention, can also lead to major calamities. As Adler notes, 'One of the most interesting things about the catastrophe at Knight Capital Group—the trading firm that lost $440 million this week—is the speed of the collapse. News reports describe the bulk of the bad trades made in less than an hour, a computer-driven descent that has the financial community once again asking if the pursuit of profits has led to software agents that are fast yet dumb and out of control' (2012).

It is easy to forget here the extent to which 'state' investments in technology and science are circulated back into the *private* investments, primary circuits and forms of accumulation. In the case of outer space, for example, government spending on the U.S. missile defence program (which, despite recent cuts, is now well above the average spent during the Cold War period) is ploughed into the primary circuits represented by corporations such as Boeing, Lockheed Martin, McDonnell Douglas and Northrop–Grumman (See Chap. 5).

Capital, the Cosmos and 'Satellite Farming'

A good example of these processes of capital accumulation and their unwelcome effects is today's industrialised form of farming. Here, we can first see some of the 'pros' and 'cons' involved in satellite-based production. Global Positioning Systems (GPS) are based on a relatively small number of satellites, and they are able to provide information about time and location to any position on Earth.

One comparatively new but now rapidly emerging example is 'satellite farming', a system based on satellite-based information about land conditions. Data on the farming land (including soil moisture and levels of organic matter) is provided to farmers. This is accompanied with advice about appropriate fertiliser applications and forms of equipment to buy and use. 'Precision' farming even includes satellite guidance of remote-controlled tractors, with farmers making real-time corrections via their on-board personal computers to remotely controlled tractors or combined harvesters. All this is conducted 'to a level of accuracy down to a centimetre'. The whole system depends on communications between large companies and farmers using electronic, satellite-based, communications, the ultimate aim of course being to increase yield, output and the enhancement of profits.

'Satellite farming' as outlined above may sound at first like a purely positive innovation, one resulting in the optimum productivity of agricultural land. There is truth in this assessment but, looked at more closely, these satellite technologies are also changing social relations on earth and not necessarily for the better. Satellite farming is nowadays characterised by an increased transfer of economic power away from farmers and towards already-powerful classes and institutions who use communications satellites to manage 'farming' at a great distance.

Very large corporations such as John Deere are promoting and providing 'precision farming' and, in the process, they are assuming a high degree of control over the whole of the farming process. And, as a result, the power, tacit skills and knowledge of farmers and their operatives are in real danger of being marginalised or even dispensed with. All as a result of making farming more 'efficient' and profitable (Addicott, 2020).

References and Further Reading

Addicott, J. E. (2020). *The Precision Farming Revolution. Global Drivers of Local Agricultural Methods.* London (?) Palgrave Macmillan.
Alizio, W. (2015). *One Satellite Surveillance System. America's Very Own Mental Illness, Paranoia.* Create Space Independent Publishing Platform.

Armitage, J., & Graham, P. (2001). Dromoeconomics: Towards a Political Economy of Speed. *Parallax, 1*, 111–123.
Austermuehle. (2016). Monitoring Your Employees Through GPS Practices. https://www.greensfelder.com/business-risk-management-blog/monitoring-your-employees-through-gps-what-is-legal-and-what-are-best-practices
Blakemore, C. (2011). *From Workplace to Social Spy: New Surveillance in (and by) the Workplace*. National Union of General and Municipal Workers. http://www.gmb.org/PDF/Surveillance
Breuer, S. (2009). The Nihilism of Speed: On the Work of Paul Virilio. In H. Rosa & W. E. Scheuerman (Eds.), *High-Speed Society*. Pennsylvania University.
Brin, D. (2009). Suggestion No. 5: Avoid a Crisis Caused by "Just-in-Time" Economics. Available: http://www.davidbrin.com/suggestion05.htm
Buschschluter, V. (2009). Satellite Helps Fight Illegal Immigration. *BBC News*. Available: http://news.bbc.co.uk/1/hi/world/europe/7818478
Castells, M. (2012). *Networks of Outrage and Hope*. Polity.
Crosthwaite, P. (2011). The Accident of Finance. In Dixon, P. (2022) *Surveillance: A Reference Handbook. Contemporary World Issues*. Amazon, Greenwood.
Dickens, P. (2009). The Cosmos as Capitalism's Outside. In D. Bell & M. Parker (Eds.), *Space Travel and Culture. From Apollo to Space Tourism*. Wiley Blackwell.
Dumenil, G., & Levy, D. (2013). *The Crisis of Neoliberalism*. Harvard University Press.
Dwayne, A., Day, J., & Logsdon, J. L. (1998). *Eye in the Sky: The Story of the Corona Spy Satellites*. Smithsonian.
Foucault, M. (1977). *Discipline and Punish*. Vintage.
Fuchs, C. (2010). Foundations of the Critique of the Political Economy of Surveillance. *The Internet & Surveillance Research Paper Series*. Edited by the Unified Theory of Information Research Group, Vienna, Austria. Available: http://www.sns3.uti.at/?page_id=24
Harvey, D. (2003). *The New Imperialism*. Oxford University Press.
Harvey, D. (2006). *The Limits to Capital* (2nd ed.). Verso.
Harvey, D. (2010). *The Enigma of Capital*. Profile.
Heires, K. (2012, September 26). Could GPS Hackers Cause the Next Flash Crash? *Institutional Investor*. http://www.institutionalinvestor.com/Article/3094717/Trading-and-Technology/Could-GPS-Hackers-Cause-the-Next-Flash-Crash
Hencke, D. (2005, June 7). Firms Tag Workers to Improve Efficiency. *The Guardian*.

Lazzarato, M. (2020). Immaterial Labour. https://www.generation-online. org/c/fcimmateriallabour3.htm
Leed, P. (2020). *The U.S. Space Force in Action*. UPDOG, U.S. Military Branches.
MacDonald, F. (2007). Anti-*Astropolitik*: Outer Space and the Orbit of Geography. *Progress in Human Geography, 31*(5), 592–615.
Marazzi, C. (2011). *The Violence of Financial Capitalism*. Semiotext(e).
Mason, P. (2013). *Why It's Kicking Off Everywhere* (2nd ed.). Verso.
Michl, T. (2011, July/August). Finance as a Class. *New Left Review*, 118–125.
Norris, P. (2008). *Spies in the Skies: Surveillance Satellites in War and Peace*. Faber and Faber.
Parks, L. (2005). *Cultures in Orbit*. Duke University Press.
Parks, L., & Schwoch, J. (Eds.). (2012). *Down to Earth. Satellite Technologies, Industries and Cultures*. Rutgers.
Petruno, T. (2013, September 14). Five Years After the Financial Crash, Many Losers – And Some Big Winners. *Los Angeles Times*.
Rangwala, G. (2011). The Arab Spring Is Not About Twitter. *Cambridge Alumni Magazine, 65*, 40–41.
Ritzer, G. (2013). *The McDonaldization of Society*. Sage.
Roberts, J., & Armitage, J. (2006). From Organization to Hypermodern Organisation. *Journal of Organizational Change Management, 19*(5), 558–577.
Shammas, V. L., & Holen, T. B. (2019). One Giant Leap for Capitalistkind: Private Enterprise in Outer Space. *Palgrave Communications*. https://doi.org/10.1057/s41599-019-0218-9
Sipri, S. (2022). *Outer Space – Battlefield of the Future?* Routledge.
Sivollella, D. (2019). *Space Mining and Manufacturing. Off-World Resources and Revolutionary Engineering Techniques*. Springer.
Slezak, M. (2013, March 2). Space Mining: The Next Gold Rush? *New Scientist*, 8.
Space Foundation. (2020). www.spacefoundation.org/media/press-releases/ spacefoundation-2020-report-reveals-122-percent-global-space-industry-growth
Sreberney, A. (2000). The Global and the Local in International Communications. In J. Curran & M. Gurevitch (Eds.), *Mass Media and Society* (4th ed.). Arnold.
Warf, B. (2013). *Global Geographies of the Internet*. Springer.
Wills, J. (2016). Satellite Surveillance and Outer-Space Capitalism: The Case of MacDonald, Dettwiler and Associates (Chap. 2). In P. Dickens & J. S. Ormrod (Eds.), *The Palgrave Handbook of Society, Culture and Outer Space*. Routledge.
Wills, J. (2017). *Tug of War. Surveillance Capitalism, Military Contracting, and the Rise of the Security State*. McGill-Queens University Press.

5

Working in 'The Silent Sphere of Production'

Conducting wars, carrying out scientific work, making communications-systems, moving money, undertaking surveillance and enabling space tourism all have one thing in common. They all make use of rockets and spacecraft. To these ends, a cosmic capitalism requires the production of rockets and satellites.

The labour processes involved in producing rockets are often forgotten. Marx referred to 'the silent sphere of production', and this particularly applies to the production of spacecraft. Very little is known about how the production of spacecraft takes place and the implications for the employees involved. So for these reasons a quite extended account of the rocket-production process is introduced here.

But before further addressing these issues, a little historical perspective is useful. Especially instructive are the early—and absolutely horrific—production lines used to make Nazi spacecraft. These were the production and assembly lines used to produce the V1s and the V2s designed to create havoc and death in London.

The labour processes taking place in these production lines were extraordinarily secretive and protected from the public gaze, and the secrecy was for a very good reason. The tunnels were used to contain literally thousands of captured Jews, gypsies and other 'minorities'. They were made to work exceptionally hard, and they were treated cruelly by the guards. Approximately three thousand people died in these labour process, some even committing suicide. And note that Werner von Braun helped to design and operate this concentration camp before he was transferred to the United States. He went on to design the much publicised Saturn rockets which carried U.S. astronauts to the Moon in the late 1960s and early 1970s.

Making Spacecraft Today

The horrors of the Nazi labour process have long gone. But the production of spacecraft often goes missing in today's accounts of 'outer space'. So the aim here is to start uncovering what Marx called in the late nineteenth century the 'silent sphere of production', pursuing his silent sphere into some of the more obscure aspects of rocket production.

Spacecraft are designed and made for a range of purposes; these including space exploration and the launch of military assets into the cosmos, as well as visiting death and destruction on supposed 'enemies'. But the labour processes involved in the production of spacecraft are often neglected and even forgotten. Again, they remain 'silent' in Marx's 'silent sphere of production' because it is here where capitalism is being made through the exploitation of labour and the realisation of profits. In this way, labour processes are at the heart of capitalism and the creation of profits. So a mysterious veil often covers labour processes, the sphere in which company owners siphon off part of the value made by their employees and retained it for themselves or for their shareholders. This is capitalism, and it applies to the making of exotic spacecraft as much as any other capitalist enterprise. A kind of unspoken 'magic' surrounds labour processes, this despite (more accurately because of) the fact that it is precisely here where capital siphons off the surplus value made by its employees to factory owners and shareholders. It also here where human beings' creative capacities are either realised or, much more often, overridden. So it is unsurprising that this sphere of social life does not receive much attention. But precisely because of their importance, their labour processes need uncovering and examining. 'Alienation' in this particular sphere of production does not seem to have prevailed.

Much the same kind of spacecraft made in the United States and Europe can be used for very different processes. They can be used to conduct surveillance and military operations, but they can also, with the aid of small landers fitted into a spacecraft's large nose-dome, be used to understand the nature of a planet or an asteroid. This chapter addresses both these purposes, starting with the production of the landers and continuing with the rockets, which are used for these purposes.

Making Landers: The Jet Propulsion Laboratory, Pasadena

The next focus of this chapter is the Jet Propulsion Laboratory (JPL) at Pasadena, California. The main aim of this organisation is to develop scientific understandings of the cosmos. They are not a capitalist organisation making profits. Rather, JPL's aim is to enhance scientific knowledge through the use of the 'landers' made by highly skilled employees.

So, labour at JPL is the starting point for making craft for important scientific missions. This means JPL machinists make the small but celebrated 'landers' which settle on the surface of the Moon or Mars (Often with considerable TV coverage). At the time of writing, JPL craftsmen are making a lander which will examine the physical makeup and possible signs of life on the Red Planet. To make such achievements, JPL employees back in California use their skills and their knowledge of materials to carry out extraordinary skilled work while making the small landers to be carried by spacecraft to Mars and to start their inspections. At the time of writing, for example, they are being directed by JPL scientists to inspect The Red Planet for signs of early life.

The Jet Propulsion Laboratory in fact has an interesting history since it was here that some of the earliest U.S. space scientists carried out some of the earliest and most influential forms of rocket production. This work was the original foundation of JPL. It was carried out in 1943 by graduate students, including Frank Malina and Jack Parsons. In 1944, the colourful Parsons was expelled from the early Jet Propulsion Laboratory because of his 'unorthodox and unsafe working methods'. These resulted in him killing himself while inventing new forms of rocket propulsion. (But it remains unclear whether this was an accident or a case of suicide.) Later, and somewhat surprisingly, JPL emerged from the shadows of World War II to link up with Werner von Braun's engineers who had arrived in the United States from Nazi Germany and were commissioned to design the Jupiter rocket based on the Nazi's V2 and which in the 1950s and 1960s launched the first U. S. 'Explorer' satellite.

More recently, JPL has been advising even Hollywood moguls, helping them to make their fantasies more 'realistic'. An example is *The Martian*, a movie made in 2015 and starring Matt Damon. Moviegoers will remember that he was accidentally left on Mars when his spacecraft was covered by a sand storm and the rest of his crew departed The Red Planet in a hurry and without Matt Damon. Fortunately, and after extended negotiations between the U.S. President and the Chinese authorities, he was rescued and brought home with the help of a large Russian space spacecraft.

The red planet of Mars in that movie was a realistic if slightly exaggerated version of the planet, one currently is being observed by a Jet Propulsion Laboratory lander.

The West Coast of the United States is where many of the original ideas of American spacecraft were invented and constructed. And with JPL making its Rovers designed to travel to the Moon and Mars, the close link between the United States' West Coast and space science have been continued and greatly reinforced. This chapter returns to JPL and its landers shortly. It is noteworthy that the kinds of libertarian politics described in Chap. 1 earlier flourished in this region of the United States.

Making Spacecraft: United Launch Alliance

At this point, I turn to the production of fairly standardised rockets used to carry payloads into the cosmos. These payloads can include either the landers described above or by the military and surveillance satellites required by government. The rockets are often made by United Launch Alliance (ULA), a very large profit-making company which makes spacecraft for a number of different projects. This includes the spacecraft which took the mysterious Space Plane into the cosmos (of which more anon).

But it also carried the JPL Mars 'Curiosity' lander to 'The Red Planet', a journey of about 40 days. The same generic design of spacecraft is made by JPL for diverse purposes, some scientific, some military.

Again, all this is a complete contrast with The Jet Propulsion Laboratory in which employees are very enthusiastic about the landers they are making. But the paradox or contradiction here is that the JPL people know little, if anything, about the large spacecraft which will carry their landers into space.

The Landers on Mars

This is where the science starts, with the landers used to start conducting scientific activity. Once close to Mars, the lander is released from the protective faring of the ULA spacecraft and it descended towards the surface of the Red Planet. The lander was then slowed down by its own small jet engines and it finally settles on the Mars surface. At this point, Earth-bound television watchers (including of course the present author) started observing the lander's progress and started sending test images back to The Jet Propulsion Laboratory. The rover unpacked its contents and locked its legs and wheels into a landing position. Then, after more checks, it started its mission of investigating the Martian surface, watched carefully by JPL employees in California. JPL scientists and its machinists have for some years been designing, making and simulating this process of landing and using their 'buggies' to investigate parts of the cosmos. At this point, it is useful to reflect on the making of the buggy.

Making a Mars Buggy: A Skilled Labour Process in Action

Constructing the small buggies means that highly skilled JPL machinists make its large number of small parts. This is what makes their work and their relations with other employees especially interesting. The machinists frequently consult JPL engineers (sometimes even during nightshifts when engineers need waking up from their sleep) making the parts that fit into the buggies. And if the parts do not fit in, they return to their lathes. Here, they adapt a part they have already made or alternatively make a new, better-fitting, part. Note that this production process is social in nature, impinging if necessary on these workers' private lives.

So JPL machinists make a large number of small and varied parts which are folded up within the small volume under the spacecraft's heat shield. Making parts of the buggies for this operation demands very high levels of skill from the JPL machinists but also, as mentioned above, close and constant interaction with other employees. These include the JPL engineers who designed the module and, on the basis of their knowledge of materials, specify the materials to be used. An interesting and important form of collaboration takes place here. On the one hand, a JPL engineer's understanding is based on knowledge in books and manuals. This means they can specify which materials should be specified for which purposes. But the machinists rely on intuitive and learned skills, these are based on experience of how different materials—including steel, brass, aluminium and copper—are likely to react when they are subjected to drilling and cutting. These skills are also politically important, in the sense that they are central to production processes as a whole.

- Once a component is made, a machinist checks with the engineers that the part fits well within the engineer's overall designs. A continuing, two-way process is now underway at JPL, with engineers proposing designs and materials and the machinists making the parts specified by the engineers. Do the parts fit together? Do they need re-making? Should alternative materials be used? Does the whole design even need a fundamental re-assessment? If a part made by the machinist still does not reach the exact specification needed, the machinist tries again. More discussion and more work ensues. If an element finally proves impossible for the machinist to make to the required accuracy, then the engineer's drawings and specifications are changed.
- The rover and lander systems require some three dozen gears and motors. Some of these open and unfold the landers to release the rovers. Others deploy the buggy's scientific and communications instruments. These are an integral part of the rover's mobility and steering system. Building these parts and combining them is hard and prolonged

work within Mangano's 'pretty ambitious schedule.' Due to the very strict constraints on size and weight, the spacecraft's rovers are compactly layered inside one another, akin to Russian dolls. The complexity of some of the parts is a result of the need to 'pack' the complex landers into the spacecraft's nose zone. These parts depend on the skill of the machinist. These people are, as one JPL machinist Keel Mangano puts it, 'configurationally challenged'. Jet Propulsion's own publicity describes the practices of Mangano and other machinists.

Box 5.1   Making a Mars Lander

It's the middle of the night at JPL, and the usual dozens of deer are on their nightly foraging rounds across the campus. Mars is up. So is the Moon. And so are nine machinists in the lab's high-precision fabrication shop, working the second shift that ends between midnight and 3 a.m. They are part of the round-the-clock team turning out odd-shaped pieces of metal that will become robots destined for Mars.

Night shift supervisor Gary Keel holds in his hand an improbable mix of geometric shapes that somehow meld smoothly together. Finely machined out a solid, 25-pound brick of titanium, the part looks like a mechanical dog's leg as dreamed up by a computer. In a way, it is. Only this is a leg for a rover of a different kind—it's a wheel strut for one of the two twin Mars Exploration Rovers now being built for launch in 2003. This part is one of thousands that will comprise the rovers. The part displays a mathematical complexity made possible only through the speedy calculations of computer-aided design. Rendering it into a real part from a drawing falls to Keel and his colleagues in the fabrication shop.

About 80% of the rover parts are of a more routine sort that will be made by machinists outside of JPL. Some parts, like the launch vehicle adapter that connects the spacecraft to the rocket, are huge—the size of a round banquet table. Others are smaller than the diameter of a pencil. Large or small, the most exotic pieces stay at JPL with staff machinists, now available 24 hours a day, 7 days a week through the crunch time of the rovers' assembly and test phase. 'We get the "Oh No!" parts', says Keel. 'The "Nobody Else Wants to Touch This" parts'. Engineers often want to keep the most complicated and challenging machining in-house, he says. That way, they can check the progress with the machinists who are bringing the new parts into being and have an opportunity to make adjustments if needed. The relationship is known as 'concurrent machining and engineering'.

(continued)

and for scientific missions with which JPL employees can positively identify. Moreover, the labour process, as we have seen, involves constant interaction with other workers. Here are their comments 'for and against' JPL. Again, they show a very positive view of their work.

Pros: 'I would say the environment is fantastic for people that are passionate about rockets and the aerospace industry, but having competing priorities outside of work starts to make the demanding environment more pronounced'.

Cons: 'Schedule pressure for the past couple of years has been top priority, and the company culture seems to have shifted to shaping the quality of work around meeting schedule objectives'.

One respondent also strongly identifies with his work as follows:

Pros: 'Captivating work, great people, great environment, and job security'.

Cons: None. 'Slow and steady advancement in business division', writes a JPL scientist. 'I have been working at NASA Jet Propulsion Laboratory full-time'.

Pros: 'A great place to work. Everybody is so friendly and eager to work with one another'.

Cons: 'None I can think of'.

Amongst JPL workers, there is a strong sense of closely identifying with, in a sense 'owning', JPL projects. This is particularly well captured by a machinist who had earlier worked on the Mars Pathfinder mission and the Cassini spacecraft which was directed towards Saturn. He states his relations to JPL as follows. 'Mars Pathfinder was my first project where I worked from the start to finish. That and the Cassini mission were the first missions I felt part of. This is the best place in the world to work. It's not the same thing day in, day out. There are challenging opportunities and a chance to advance your career. And this is the only place I know of where you work on a part for something and see it on the front page of the *Los Angeles Times* making history a few years later'.

It is no coincidence that The International Association of Machinists is one of the largest and most powerful unions in the United States. It is precisely because their skills, such as those possessed by JPL workers, bring them into key positions in the labour process and also key positions in American politics.

Realising the Rover's Potential

Finally, we must consider how the rovers are actually used, once they land on Mars. This matter has been given a remarkable degree of attention by Janet Vertesi (2015). She describes another set of labour processes which are central to the rover's purpose. This includes engineers and others at JPL first needing to be sure that their robot is safe on the harsh and demanding climate of Mars. But that is just a start of a series of processes on Planet Mars which have been planned for on Earth. As Vertesi points out, the rovers may appear to be alone on The Red Planet but they are commanded at a distance of several million miles by a team of scientists and engineers back on Earth. 'Science' in this sense is a social production. It is carried out by specialist workers who have planned the mission and decided what a Rover does (and what images it produces) back on Earth.

Vertesi adds importantly to this point, demonstrating the ways in which the rovers make images, which are shared by the scientific community and thoroughly investigated to make a collaborative understanding.

'In their myriad interpretations and projections, Rover team members employ images as a resource to both conduct their science and manage their community. Conducted with materials ready to hand and with robots millions of miles away, the work is at the same time practical, technical, social and epistemological as it makes Mars available for interaction. This team and its interactional and organisational work with image making and interpretation is particularly oriented toward the continual production of consensus, hierarchical flattening and a concomitant social order' (2015: 242). These JPL specialists remain very active once the lander has made contact with Mars. In this sense, an image taken by a Rover is a 'social production', a product of a group of scientists remaining on Earth. Furthermore, it is an end result product which requires consensus amongst these scientists. As Vertesi puts it,

> The rovers may be alone of the Red Planet, but they are commanded at a distance of millions of miles by a team of scientists and engineers on Earth, who together make decisions about where the robots should drive next and that they should do. (2015: 160)

Preparation and the 'Science' Question

A great deal of earlier work has planned the Rover's tasks, what it will image and how, perhaps most importantly, once it has landed. The well-known images of Mars, for example, are a product of extensive earlier discussions and debate between the scientists and engineers. What images should be taken? How will we interpret them? Referring to the now famous images of a lander on Mars, Vertesi shows the extent to which background work by scientists has been taken back at NASA. In this sense, the images taken are 'social constructions' reflecting not only the decisions made by a group of scientists but also reflecting their interpretations once the lander has landed and starts to make observations. Vertesi makes clear the extent to which the landers and their results are a product of social relations and processes made on Earth.

> It is extraordinary to witness an image whose representational quality is judged on how well it represents the group that constructs it, not only the object it purports to represent. While we tend to think or representations as standing between an observer and the world, they also represent an observer's work in the world. (2015: 160)

The knowledge of the space scientists dealing with the Mars lander is 'scientific' in a very specific way. It involves, for example, knowing about the lander and how it will interact with the surface of Mars. This means, Vertesi tells us, first parking the Rover on a nearby slope, this allowing the Rover's solar panels to get charged up. But this turns out to be easier said than done. The rover's wheels get trapped and there is a further complication. The rover's right wheel gets jammed, as it later turns out permanently. The team at JPL eventually extricate the Rover from these hazards and they park it in a new, hopefully safer, place. All these labour processes result in a Rover made in California achieving its original purpose of collecting information about The Red Planet. But the point here is the difference between, on the one hand, the kinds of scientific knowledge which is real and informs our understanding of cosmos from the more contingent and practical kinds of understanding about how the lander operates under the particular, 'local', conditions of Mars. Both kinds of understanding are needed to define an understand the lander and its predicaments.

Making the United Launch Alliance Spacecraft

I now turn away from Pasadena-based labour processes, by way of a complete contrast, towards the production of the spacecraft made at United Launch Alliance. Here, it will be remembered, is where the massive rockets for carrying the buggies into the cosmos are made. Note in particular, the big contrast between the labour processes making spacecraft here at ULA compared with those at JPL. In the ULA case, the employees are more disengaged, thinking of it as a 'normal' alienated job, one in which there is little or no interest in who will own the rocket and the purposes for which it will be made. Though one respondent was concerned with how ULA 'dances on the line with politics too much' (Perhaps, a reference to ULA's production of spacecraft for military purposes).

By far the most common material used for making a large spacecraft of this kind is aluminium and to a very large extent the labour processes at ULA are organised around this material. Aluminium can be readily shaped and curved into the barrel-like form, characteristic of spacecraft.

So at the start of a spacecraft assembly line are flat sheets of aluminium waiting to be further thinned down, bent and curved so that it resembles a barrel, which is characteristic of a spacecraft's body. Work on the flat plates of aluminium now proceeds. First, they are examined for the smallest of flaws by sharp-eyed labourers kneeling on the aluminium. But this is just the start of a much longer process. Next, the aluminium is milled down to one-third of its original thickness, but still leaving the material with about two-thirds of its original strength.

But next, a more intensive process takes place, with the thin aluminium sheets being subjected to computer numerical control (CNC) which bends the flat aluminium into three-dimensional shapes such as rectangles or triangles. This is partly automated process but it also demands close attention from machine operators. The end result is an even further strengthened but still light aluminium sheet.

The production process further continues. Making the aluminium into a spacecraft is carried out by immensely powerful steel presses, with each press and its steel plates being regularly monitored and checked for abnormalities. This is conducted by seated employees carefully regulating the speed of the presses as the aluminium is curved. At the same time, these employees are also looking out for possible weaknesses within the aluminium pieces, sometimes stopping the procedure to visually assess possible problems.

Here are some typical comments from workers at the ULA factory, dated 2021. (A '9/80 work schedule' consists of a total of eight nine-hour days, one eight-hour day, and one day off spread over a two-week period.)

Pros: 'I welcome the work and sense of fulfilment that goes with it, Salary, Benefits, Flex Time, 9/80 schedule'.

Cons: 'If you are wanting to be in upper management, be an engineer. Mandatory unpaid overtime for no purpose, management whacking key individuals but leaving behind scrubs'. ('Scrubs' are protective clothing wear).

Pros: The 9/80 Schedule is nice and the majority of the people are very nice.

Cons: Not enough work and too many employees in certain organisations within the company. The best projects and most interesting work is given to people that Leadership favours and they are then rewarded for it. Everyone else is left to fend for themselves. Leadership won't make any statements on anything related to the civil unrest happening nationwide because they 'don't specialize in social justice' and would be 'bad at it'. They simply ignore it and act like it's not a problem. The vast majority of the employee population is white male.

Mechanical Engineer in Vandenberg AFB, CA.

Pros: A cool industry, working on aerospace missions. Decent pay, nice people Good. Recommend. **Cons:** Not many west coast (spacecraft) launches.

Current Employee, more than 3 years.

Pros: Major Development of Vulcan/Centaur V provides a lot of technical challenge and variety. Pay, benefits and general work/life balance are great.

Cons: It is very rare for someone to get fired, making it not uncommon for underperforming individuals to be highly visible within the company. Focus on 100% Mission Success drives a lot of red tape that can be frustrating to navigate. Wide experience gap; very few mid-career

employees. Most employees work for less than 7 yrs. or for more than 25 yrs. Staffing levels don't seem to reflect the workload felt during the current push to a first Vulcan launch. (The Vulcan craft mentioned here was made to meet the demands of the U.S. National Security Space Program and is now used by the U.S. Space Force.) **4.0.**
Great Company, Dances on the line with Politics too Much.
Pros: Good Benefits and flexible schedule. **Cons:** Driven by unpaid overtime. Workload is excessive.
Current Employee, more than 3 years. Not a bad place to work.
Mechanical Engineer II in Decatur, Alabama.

Labour Processes: A Comparative Assessment

How can the above comments be summed up? This analysis of different labour processes (needed for different purposes) shows that the ULA workers are not profoundly unhappy in their work. On the other hand, there is not the same sense of collaboration and pride in skills usages as described by the JPL workers. The JPL workers also enjoy their relations with other workers and, most importantly, they know they are using their skills towards contributing to a scientifically valuable project (and one they could later watch on TV as their buggy starts trundling over the surface of Mars).

Making a Mars Lander: Conclusions

Labour processes often entail close and intense surveillance by management, this being the way in which labour power is harnessed in the interest of capital accumulation. But this chapter shows these processes and relations can leave a good deal of autonomy for the employee. Highly skilled workers such as those making a Mars Lander are not subjected to this kind of close and intensive control. The elements of a spacecraft such as the JPL Lander are made in workshops where the over-riding priority is to combine the machinists' skills with those of the engineers to make a spacecraft capable of encountering the extreme conditions encountered in outer space. And the 'normal' process of close workero surveillance clearly does not apply at the Jet Propulsion Laboratory and here, as described, greater levels of employee satisfaction prevail.

Financing Rocket Production: From Government to Private Capital

This chapter has focussed on the labour processes involved in spacecraft production. It has shown that the labour processes in The Jet Propulsion Laboratory are more human-centred and humane than those in the private sector. This is a result of making something scientifically valuable and working closely with others around a shared and worthwhile project with which they can readily identify. They can even watch 'their' product on TV, trundling around the surface of Mars.

But we need to finish this chapter on spacecraft production on a sombre note, one looking ahead to developments in the financing of rocket companies. The following headlines are unsettling for the people and the labour processes describe above.

> US Space Firm partners with Venture Capital Firm *Embedded Ventures* under new R + D Agreement. (https://technocrunch.com. 10.13.21)

Future space programmes will, it seems, be increasingly, financed by private capital and less by funds derived from taxation. This is of course part of a wider process. Capital will, as far as possible, reduce the costs of labour by de-skilling employees such as those working for the abovesaid JPL case.

Even more concerning, as discussed in the next chapter and later parts of this study, is the fact that capital is beginning to assume full control over what used to be 'government' or 'state' space projects. Whereas, for example, the spacecraft used for the first Apollo mission were made 'in house' by NASA. The spacecraft for the second mission will be powered by a private company, Northrop Grumman. In this sense outer space is being increasingly capitalised. And employees such as those working at JPL will find themselves coming under the pressure of private capital. And this will almost certainly mean that people and their skills will as far as possible be *de*skilled. In turn, it will be further invested in the production of military spacecraft. Cosmic Capitalism does not just mean capital being invested in the cosmos. It means capital is organising and paying for the production of spacecraft and that the skills of employees will no longer be recognised and paid for.

So the long-term prospects for JPL do not look particularly good. On the one hand, the pressure will be on the likes of JPL to deskill their employees as far as possible. (Though it is difficult to see how this can be achieved in the case of making objects such as the Mars Lander.) At the same time, private capital combined with government will presumably invent—and invest—in more military projects such as Space Force. These are largely financed by private capital to 'Keep America Safe' and no U.S. government is likely to reject such a popular rallying cry. But the long-term prospects for employees at JPL will for sure come under pressure, particularly if their skills are over-ridden and their labour power is sold to the lowest bidder.

References and Further Reading

Alamalhodaei, A. (2021). U.S. Space Force Partners with *Embedded Adventures* Under New R + D Agreement. See: https://techcrunch.com/2021/10/13/u-s-space-force-partners-with-vc-firm-embedded-ventures-under-new-rd-agreement/

Braverman, H. (1998). *Labor and Monopoly Capital. The Degradation of Work in the Twentieth Century*. Monthly Review Press.

Dickens, P. (2020). Labour Processes, Skills and Outer Space. In *Work in The Global Economy*. University of Bristol.

Thompson, P., & Smith, C. (2010). *Working Life. Renewing Labour Process Analysis*. Palgrave.

Tsvgkh, V., & Netipa, D. (2019). The Role of Trade Unions in the Development of the Aerospace Industry. https://www.researchgate.net/publication/33627 1037_The_Role_of_Trade_Unions_in_the_Development_of_the_Aerospace_Industrypdf

V2 ROCKET.COM – rocket.com/start/chapters/mittel.html

Vertesi, J. (2015). *Seeing Like a Rover. How Robots, Teams and Images Craft Knowledge of Mars*. University of Chicago Press.

Wills, J. (2016). Satellite Surveillance and Outer-Space Capitalism. In P. Dickens & J. Ormrod (Eds.), *The Palgrave Handbook of Society, Culture and Outer Space*. Palgrave.

6

Cosmic Capitalism and the Body

The astronaut's body does not nowadays receive much popular attention. Perhaps, this is because she or he is covered by protective clothing, making the astronaut anonymous. Gone are the days of *Dan Dare* in the *Eagle* comic, with our hero somehow not needing a helmet, even in the distant reaches of the cosmos. But this absence of headgear made him into a handsome hero, one immediately recognisable to the mass cosmic-reading public.

This chapter gives extensive study to the human body in the cosmos. Here are some of the most more important matters arising from a cosmic form of capitalism. Most obviously, these include the lack of gravity, a lack of breathable air and the changing clashes between the body and the rhythms of the cosmos. How, if at all, do these relate to a body which evolved on Earth? And considering that the cosmos is now being made into a warfighting zone, how will military personnel remain effective under the trying and testing circumstances imposed by the cosmos?

This chapter is an extended version of Dickens, P. (2019) 'Social Relations, Space Travel and the Body of the Astronaut' in Cohen, E., Spector, S. *Space Tourism. The Elusive Dream.* Bingley, Emerald.

But there is a more fundamental question. It has for some time been assumed that the body will make ever-increasing incursions into the cosmos. But this may not necessarily be the case. As mentioned in the last chapter and in subsequent chapters, it seems quite likely that 'long-haul' missions into the far reaches of the cosmos may be quite uncommon. Exploratory scientific missions of this kind may well continue, but, for the foreseeable future, capital will not be spreading into the far distant regions of outer space. This is because capital, along with its investments nearer to Earth (particularly in the military) is making good profits. There is no immediate need for capital to start accessing the far distant reaches of the cosmos and start extracting for 'rare Earths'. There remain such rare Earths in Earth itself and capital is unlikely to start serious extraction elsewhere.

Nazi Space Medicine: The Objectified Body

What is now called 'space medicine' has shocking origins. And, as later discussed in this chapter, there are serious lessons here for today. Today's type of 'space medicine' was originally conceived and practiced by the Nazi regime during World War II. (Jacobsen, 2014). The stated objective was to understand how the body reacted to the extreme conditions of the upper atmosphere.

But 'space medicine' in this case had further, extremely sinister, objectives. Nazi 'doctors' threw Jews, gypsies and other 'subnormal' people from planes into the upper regions of the atmosphere. They did this on the entirely wicked pretext that the victims were 'sub-human' and could therefore be used to 'scientifically' discover how the human body fared in the atmosphere. Jews and others were in effect made into 'animals' available for 'testing'.

'Space medicine' in the Nazi case consisted of direct and complete control over the human body in space. But of course it was a thinly disguised way of construing human beings as 'animals' which could, like non-human 'animals', be legitimately experimented on and killed. And under the Nazi regime other supposedly 'inferior' people included not only Jews but also gypsies and 'mentally disturbed' individuals. They were imagined as subhuman people or animals which could be legitimated and subjected to exceptionally cruel practices. They were categorised as people with 'low intelligence' who could be legitimately killed. The long-term aim of such experiments was of course to make a new 'super-race'. And to this end, Jews and others were treated not as human beings but as 'animals' who could be killed as non-members of the human race. There are obviously cross-overs here with the human 'animals' used to make Nazi rockets.

Science and the Body After the Nazi Experiments

Following the defeat of the Nazis at the end of World War II, a selected group of German (and Nazi) scientists were captured and transferred to the United States (Jacobsen, 2014). They were used to counter the supposed threat of the Soviet Union government which, it was argued, was planning to attack the United States, following the 'The Cold War'. The scientists extracted from Germany included Dr Hubertus Strughold who had been working at the Medical Research Institute, Berlin. He had presided over some of the Nazi killings but denied any such action.

As World War II came to an end, Strughold and over 1600 other senior German scientists were swiftly moved to the United States. Their leader was Wernher von Braun. The main aim of the U.S. authorities was to stop the Soviet Union from capturing these scientists and using their knowledge of space weaponry and nuclear power to make the Soviet Union into a dominant military power. Instead, they were 'recruited' by the United States and NASA. The American military and its politicians 'went to great lengths to whitewash his dubious past' (Jacobsen, 2014: 2). Any talk of Jews used as animals for spaceflight now of course evaporated.

In 1948, Strughold arranged for a rhesus monkey to be fired into space in a rocket modelled on the Nazi V2. The objective was actually reminiscent of the earlier Nazi killings, but now live monkeys rather than people were tested under the extreme conditions in the cosmos. The monkey in the V2 inevitably died but it is still seen by some space historians as a valuable first step towards the wider 'success' of the U.S. authorities in sending 'manned' rockets into the cosmos and later to the Moon.

In 1961, Strughold claimed to a U.S. reporter that he had never had knowledge of killing Jews and others in Nazi Germany. The only people in his Berlin institute who knew about the killings were, Strughold said, the janitor and the man who took care of the animals. And it was only these people, he said, who had been members of the Nazi Party. The U.S. authorities accepted his position and later celebrated him as 'The Father of Space Medicine'. In these ways, 'Space Medicine', with its early horrific use of objectified human and animal bodies, was introduced into the so-called space race between the United States and the Soviet Union.

Since the beginning of space travel, a total of eighteen astronauts and Russian cosmonauts have died during the space flight missions. This seems a relatively low number but it is in large part a tribute to the people in Nazi Germany and the animals sacrificed in early U.S. rocketry who had been made to pave the way for manned space missions.

The Soviets, the Animal's Body and the Cosmos

Meanwhile, animals were chosen by Soviet officials to assess the viability of spaceflight for human beings. The animals included a husky dog in possession of the Soviet government which had been found roaming the streets of Moscow. Her keeper became attached to her and gave her the name Laika. Naming the dog in this way in effect signified the dog's elevation to 'honorary human' status.

An image of Laika also appears on a giant Obelisk in central Moscow. It is entitled Monument to the Conquerors of Space. It reads: 'This monument was constructed to manifest the outstanding achievements of the Soviet people in space exploration'. In this way, Laika was categorised as one of 'the people'.

'Astronauts on Strike': The Body and the Cosmos Today

This chapter now turns to the human body's relations to the cosmos in our own times. It starts by examining in a relatively 'objective' way, the management of astronauts' bodies by U.S. mission controllers. This has parallels with the control of labour by capital on Earth. It finishes by stressing some these astronauts' subjective experience.

Skylab and Astronaut Resistance

The issue of control and autonomy for astronauts arose in 1974 with one of the Skylab missions designed to undertake observations of the Sun and the Moon. Previous astronauts on this programme had been unable to collect all the data required of their mission. A catch-up was needed for the whole Skylab programme to be deemed a success. As a result, the ground-based controllers demanded an intensive 16 hour/day for the final part of the Skylab mission. This required intensive control of the crew. It consisted of minute-by-minute work schedules for the three astronauts during their 84-day mission.

But the astronauts went on strike against the work intensification imposed upon them. This was similar to struggles between managers and employees in the workplace but now transposed to the cosmos. The SkyLab4 astronauts firmly resisted their managers' demands, saying they would never have been expected to work for 16 hours a day for 84 consecutive days on the ground. They continued in outer space cleaning up after their experiments. But soon after landing on Earth, they were sacked. The whole event raises issues of objective power and subjective experience associated with rocketry and outer space. This chapter focuses on these connections.

Management, Labour Relations and Power

A recent study of astronaut–management relations on the International Space Station documented some of the intensive management–astronaut relations now involved. Astronauts were in this case reduced to little more than automated technicians. The pace and rhythms of the day are unequivocally set by managers on the ground. The cyclical rhythms of the body in outer space were made constantly subject to monitoring and control by supervisors on Earth. To this end, ground controllers now attempt to impose a 'linear' regime on the astronauts, one in which standardised tasks must be undertaken as part of an organised, regular sequence.

An 'Onboard Short Term Plan Viewer', or OSTPV, shows an electronic display of mission plans for individual astronauts. And all this information is shared by mission controllers, astronauts and a range of control stations located across the Earth. Across the top is a timeline showing the days and nights of a particular mission. Of particular interest to this chapter is the list of astronauts listed down the left-hand side of the diagram and their planned activities listed across. As and when their missions are changed, the diagram is changed. Importantly, all this information is shared by astronauts, their managers, scientists and technicians across the globe. Foucault, drawing on the work of William Bentham in the eighteenth century, described modern society as managed in an invisible way, with prisoners subjected to constant surveillance by authority.

So the pace and rhythms of the astronauts in this case are unequivocally set by Mission Control. Here, the attempt is made to reduce astronauts into little more than automated technicians reporting back to Mission Control. And to this end, life on the station is managed via spreadsheet, with every minute of each astronaut's workday being mapped out in blocks devoted to specific tasks. And when an astronaut clicks onto a time block, it is expanded to show all the steps necessary to perform the task at hand—whether it is conducting a long process such as extinguishing a fire in the spacecraft or perhaps simply stowing supplies from a recently arrived cargo ship (Fishman, 2015: 15–16).

So here we have the astronaut's body conceived as an object, one controlled by a 'subject' based back at Mission Control and their access to the time blocks. In this respect, the astronaut's body remains highly managed.

Controllers, Power and Authority

But note also the tensions here between controllers' demands, on the one hand, and the astronauts' experience, on the other, the former exerting control and the latter attempting to gain higher degrees of autonomy and flexibility.

In this latter respect, note that astronauts often made comments on the rigid linear order imposed by their superiors. One astronaut complained, for example, of being given 'only 30 minutes [scheduled] to execute a 55-step procedure that required collecting 21 items. It took 3 or 4 hours'. Another astronaut records in his journal that 'it has been a pretty tedious week with tasks that were clearly allotted too little time on the schedule. Talking to [a Mission Control staff member] today, I realized he just doesn't understand how we work up here'.

Yet astronauts also find ways to resist the controllers' authority. Note here Lefebvre's concept of 'appropriated time'. This concept has a special significance for understanding the astronaut's body in the cosmos (although Lefebvre himself was an urban sociologist and had little interest in outer space). 'Appropriated time' is in Lefebvre's words, 'time that forgets time, during which time no longer counts (and is no longer counted)'. The reflections of Marc Garneau, the first Canadian to gain access to the cosmos, offers an example of 'appropriated time', one carved out in opposition to the highly organized 'linear' control over time directed by the managers from Earth.

Garneau wrote,

What you aren't ready for being the first time in space—on an emotional and intellectual level—is how looking down at Earth will profoundly affect you. Over the long term, it has changed the way I think about planet Earth. When you go around the planet and look down, you think about the fact that this is the cradle of humanity, that this is a place where seven billion people, 200 countries, live side by side, that we share this place and there's nowhere else to go.

Cyclical Rhythms, the Body and Control

Recent empirical research shows that the interactions between body rhythms and outer spatial rhythms are far more complex and far-reaching than simulations and convenient assumptions would suggest (Barger et al., 2014). But an understanding of these interactions remains important for successful missions. In the case of space shuttle missions and the months spent on the International Space Station, for example, astronauts were left struggling to get adequate sleep and rest. The fact that the sun might seem to 'rise' into view every 90 minutes. As such, it is a good indicator of the difficult tests to which the human body is subjected. How, in short, do controllers manage their astronauts when 'days' for the astronauts are reduced to only 1.5 hours?

Such irregular clashes between light and dark also make it very difficult for astronauts to remain alert and efficient. A study of sixty-four astronauts on eighty shuttle missions and twenty-one astronauts on International Space Station missions showed, for example, that 'sleep deficiency is pervasive among crew members' (Barger et al., 2014: 910). Astronauts slept for just 6 hours per night on average, when mission controllers were advising 8.5 hours. Astronauts have often turned to sleeping pills to compensate for such conflicting and changing relations between the body and cosmos stemming. About 78 per cent of shuttle mission astronauts have used sleeping pills to regulate their body's 'days' and 'nights' (Barger et al., 2014, op.cit.).

But of course sleeping pills are hardly a viable solution to the changing clashes between the rhythms of the astronauts' bodies and those of the cosmos. This is especially the case for longer missions, since astronauts on such missions can only get a few hours of sleep at a time and a 'sleeping pills solution' could adversely affect their performance and even endanger their judgement. Some of the research on this topic is reported in a journal named *Military Medical Research*. The implication is that tiredness and the use of sleeping pills in military projects may well not be an ideal combination for people involved in military engagements with possible enemies in space. It must be said, however, this kind of Star Wars engagement nowadays seems very unlikely to take place. This seems another way in which thinking about humanity in outer space has taken a 'fantastic' form, one detached from material reality.

Complex conflicts between the rhythms of the cosmos and the internal rhythms of the body persist without any obvious 'solutions' beyond taking yet more sleeping pills. And as a result of space travel being increasingly composed of longer missions, relations between astronauts and their controllers are now changing. Though crews on extended missions into the cosmos still find their 'days' and 'nights' closely managed by NASA Mission Control.

Remember too that astronauts, their movements and decisions in outer space, remain closely monitored by Mission Control, especially when their internal body clocks no longer coincide with the rhythms of Earth and other celestial bodies. Here is another source of potential destabilisation for the human body. The rhythms of the circuits of the Earth, the Sun, the Moon and Mars at this point could have imposed on the rhythms of the astronaut's body. And this is where some regulation is needed, even on shorter missions including those on the International Space Station. To what extent can the body really be made to 'adjust' in these ways to the demands of their controllers? And some heroic assumptions have to be made by controllers regarding the 'flexibility' of the astronauts' body as it combines with the changing rhythms of the universe. And this links to a key issue concerning space travel if it increasingly penetrates the far reaches of the cosmos. Crews on extended missions do not usually find their experience to be boring and monotonous. But this is mainly because their activity remains intensively managed by NASA's Mission Control. Their aim is to keep the astronauts alert and busy. When, for example, one of the Skylab crew appeared insufficiently occupied by their assigned working hours, Ground Control quickly found more tasks for them to undertake (Peldszus et al., 2014).

Note that the cyclical rhythms of the body in outer space are constantly subjected by mission controllers on Earth. These controllers attempt to impose a 'linear' regime on the astronauts, one in which standardised tasks must be undertaken as part of a regular sequence. But this can cause tension between the controllers and astronauts, since the latter would prefer not an enforced regime but a high degree of autonomy and flexibility for their labour processes.

In fact astronauts often criticise the rigid linear order imposed by their 'superiors', a system not unlike that experienced by employees engaged in Earth-based labour processes. One astronaut on the ISS complains, for example, of being given 'only 30 minutes [scheduled] to execute a 55-step procedure that required collecting 21 items. It took 3 or 4 hours'. Another astronaut recorded in his journal: 'It has been a pretty tedious week with tasks that were clearly allotted too little time on the schedule. Talking to [a Mission Control staff member] today, I realized he just doesn't understand how we work up here'. These comments could have been easily made in a workplace back on Earth.

And, as the astronauts' 'strike' described earlier suggested, astronauts also find ways to resist the controllers' authority. The French sociologist Henri Lefebvre introduced the notion of 'appropriated time' when considering the relationship between power and human agency. This concept has a special significance when considering the body in the cosmos. 'Appropriated time' is, in Lefebvre's words, 'time that forgets time, during which time no longer counts (and is no longer counted)'. The reflections of Marc Garneau, quoted above, offer an excellent example of 'appropriated time', one in opposition to the highly organised 'linear' control over time directed by the managers from Earth.

Threats to the Body in the Cosmos

What are implications for the human body inserted within these differing and intersecting types of rhythm? On Earth, bodily rhythms coordinate relatively well with each other, resulting in a relatively stable state of health. And again, these bodily rhythms are linked to cyclical movements such as day and night while also having their own internal cadences determined by millennia of physiological evolution.

And research into human beings' health in space runs the risk of conceiving astronauts and their heath as technical problems to be solved. But a broader picture of the body and the cosmos is needed and elements of such a perspective are again provided by Henri Lefebvre. He had no specific interest in the body and outer space but his work on the rhythms of the body and of daily life are uniquely helpful for further understanding the often complex relations between astronaut's body and outer space.

Lefebvre's form of Marxism, and his notion of 'rhythmanalysis', helps us to recognise and examine the intersecting and often conflicting relations between the rhythms of the universe, the rhythms of the body, and the regimes attempting ground-based mission control over astronauts. 'Mission control' in the form of Earth-bound managers and controllers regulates astronauts' life in the cosmos. It has ready access to large amounts of complex information on the body but its changes in space which are simply and understandably unavailable to the astronauts.

This of course means great power accrues to Mission Control. Lefebvre' work puts this authority in context. He suggests that the rhythms of the body in outer space should not be considered as simply impositions, the product of power hierarchies in organisations such as Mission Control. They also have an acting role, one stimulating a person's imagination of the body and in our case the experience of the circuits and rhythms of the universe itself.

So Lefebvre's 'rhythmanalysis' can be extended to examine not only the conflicts between the bodily rhythms which have evolved on Earth and the rhythms of the planets as they circle the Sun. These are real enough but Lefebvre's vision also focuses, most importantly, on experience. In our case, the changing experience of astronauts as they progress into the cosmos. A Lefebvrian vision alludes to, and neatly sums up, the position of the astronauts, and most importantly, the experience of the cosmos. 'The cyclical originates' Lefebvre writes 'in the cosmic, in nature: days, nights, seasons, the waves and tides of the sea, monthly cycles, etcetera. …. Great cyclical rhythms last for a period and restart: always new, often superb, inaugurates the return of the everyday' (1992: 8). In short, Lefebvre offers to less than a way of understanding the human body's complex relations with the cosmos.

So Lefebvre's account, one alluding, on the one hand, to changing clashes between different types of days and nights and, on the one hand, and the rhythms of the cosmos, on the other, suggests some extraordinary future experiences for not only by astronauts but in future many human beings. They may be subject to Mission Control but at the same time they are also capable of appreciating their experience and acting accordingly.

Risk and the Body

What are implications for the human body inserted within these differing and intersecting types of rhythm? On Earth, bodily rhythms coordinate well with each other, resulting in a relatively stable state of health. And again, these bodily rhythms are linked to cyclical movements such as day and night while also having their own internal cadence, determined by millennia of physiological evolution. But for astronauts circling Earth and undertaking long-distance space travel, these body rhythms do not necessarily combine well at all with the cosmos they are experiencing. The astronaut's body has evolved from its relations with the alternating phases of light and dark, day and night. But in space, it is subjected to frequently changing relations to the cosmos with no obvious 'days' and 'nights'. Little research has been conducted on the effects of these contrasting and changing rhythms on the human body. Yet, as with the problem of gravity levels, the prospect of long-distance space travel makes this area of research both interesting and urgent. For example, astronauts in the space shuttle missions now struggle to get adequate sleep and rest. For them, the 'morning Sun "rises"' every ninety minutes. This may be novel and exciting but for the human body it is a nightmare, one it struggles to accommodate. One study in a journal named *Military and Medical Research* even draws out some possible implications. Astronauts, the study suggests, will be so groggy as a result of these changing relations between the body and the cosmos that they will not even be able to adequately fight their cosmic enemies. But, many would add, if such contradictions afflicting the body result in stopping intergalactic warfare, the supposed 'problem' may not be such a bad thing.

Threats to the Body: Radiation

Of particular concern nowadays is the effect of radiation exposure on the human body. Phillips addresses this question directly as follows.

'Based on our current knowledge, Mars voyagers will be exposed to a level of radiation that would not be acceptable for nuclear power plant workers or hospital x-ray technicians. One prediction by a NASA radiation expert in 2004 was that the added risk of cancer from a 1,000-day trip to Mars and back was somewhere between 1 percent and 19 percent for a healthy 40-year-old male, and somewhat greater for a female because of the possibility of breast and ovarian cancers'. So the conclusion here is that the human body is meeting its serious limits in the cosmos. But why is this necessarily a bad result?

Similarly, other impacts on the human body may have negative impacts. NASA and the rapidly growing private space industry are now much less concerned with catastrophic accidents and explosions on or near launch sites. Instead, they are now much more concerned with emerging challenges to the body over long-distance missions.

A range of problems regarding the body has been documented, and they are taken very seriously by NASA. Spacecraft requires human skills to monitor and cope with changing conditions during a protracted stay in outer space. But again, the frailties of the human body could surely be celebrated in such a context. Furthermore, there is no particularly good reason why humanity should spread further into the cosmos. Imaginaries about humanity extending to the cosmos have again got out of sync with a more boring reality.

A similar conclusion applies to animals. And this means, tests and results based on small animals are of very limited value. Small animals such as rabbits and mice continue to be used to test the effects of microgravity on muscle loss and skeletal change. But again, they are of limited value to the success of today's long-distance rocket programme, and if more animals have their lives saved as a result, this is no bad thing.

Another Threat: Spacesuit Failure

A further and more obvious threat to human body in the cosmos arises if the protection offered by a spacesuit or spaceship for some reason fails. Phillips (2014a) predicts a dramatic and very painful death if this should happen. His warning is couched in suitably catastrophic terms.

> The space environment is very unfriendly. It is an almost complete vacuum that would cause you to essentially explode if exposed without the protective atmosphere inside a space suit or spaceship. Actually you wouldn't explode, but all of your body fluids would begin to boil due to atmospheric pressure, and you are mostly made of water. This would be a rapid and unpleasant end.

NASA's Human Research Program is now re-examining the effects of artificial gravity on the body during long-duration flights. As outlined earlier, convenient and often inhumane assumptions were made about the links between the body and the cosmos in the earliest days of humans in rockets. But now, and assuming astronauts will travel over even longer missions and in the process constantly monitored from Mission Control, the human body in space is receiving much more serious scientific attention, particularly by NASA. The ethical issues surrounding the body place a huge query over such a massive and long-term project. But it is not clear why, if the human body should be seriously limited to Earth, this should necessarily be a poor result.

From Alienation to Autonomy?

Space travel nowadays involves increasingly long missions. And relations between astronauts and their controllers are changing. Yet, so far at least, crews on extended missions have generally not found their experience boring and monotonous and this is in part because, as discussed earlier, their every activity is intensively managed by NASA ground control. NASA aims to keep the astronauts busy by imposing heavier individual workloads. When, for example, the aforementioned Skylab crew appeared insufficiently occupied by their assigned working hours, ground control quickly found more tasks for them to work on.

But as the duration and distance of missions into outer space are further extended, ground controllers have been obliged to allow higher degrees of agency amongst astronauts. Key to this effort is the simulation of terrestrial activities. One study even described NASA's recent long-duration missions provided what I called a 'microcosm of home life'. The simulation of home life in the cosmos—via of course satellite communications—apparently allows astronauts to engage in special occasions such as holidays, family birthdays and especially notable football games. In these ways, the links between the astronaut's body and Earth's rhythms is, supposedly, at least restored. (Though if the football team followed by a couped-up astronaut keeps losing, she or he might well stop contact with Earth.)

Days, Nights and Wars

Much of the work concerning the rhythms of the body in space is conducted by the military in the United States and China. And, significantly, it is published by military journals such as *Military Medical Research*. The expansionary mind-set here remains and again there is no sensible reason for the military, and its back-up space industry, to engage in such projects. Similarly, as mentioned earlier, space scientists seem concerned with clashes between the rhythms of the body and the rhythms of the universe. And they recommend sleeping pills to enforce a particular regime on the astronauts' body. But this issue again falls away if the need for such expansion, and the expansion of capital to this end, is abandoned. The world will not, to coin a phrase, come to an end.

Another Problem: Returning to Earth

Re-engaging with society on return from the cosmos brings both rewards and dangers. Smith's 2005 study of the nine surviving lunar astronauts is very suggestive. All these moon-walkers struggled to adjust to the social regimes and rhythms they faced when returning to Earth. This was partly due to the 'overview' effects of perceiving Earth and humanity from such a literal and figurative distance, and the resulting epiphanies that many of the moon walkers experience in the cosmos. For example, Marc Garneau, the first Canadian to enter outer space, reported that 'there are wars going on, there's pollution down there, but these are not visible from up above. It just looks like a very beautiful planet'.

The experience of circling Earth also leads to some astronauts developing new conceptions of not only of the cosmos but also of themselves and their priorities. This so-called overview effect has even been given physical recognition in the design of the International Space Station. NASA now provides a special 'cupola' enabling cosmic meditations by astronauts. And by all accounts, it is regularly used by International Space Station astronauts.

After such intense experiences in the cosmos, returning to the workaday rhythms and routines of life back on Earth can cause astronauts to be a mentally destabilising stress. For example, Neil Armstrong was the first man to walk on the Moon. But he virtually disappeared from public life on his return. In fact he became notorious for his absence, even failing to turn up to dinners and other events held in his honour. Meanwhile, Alan Bean, an astronaut on the Apollo 14, became increasingly interested in the paranormal and the idea of 'an Intelligence in the Universe'. He believed he had glimpsed such an intelligence on the Moon and, on his return, he founded an Institute of Noetic Sciences to undertake scientific research into the subjective experience of the cosmos.

Alan Bean (Apollo 12) also developed a passion for painting on his return to Earth. But his only subject was the surface of the Moon. On his return to Earth, Bean seemed to view human life with the bemused wonder of a visiting alien. Andrew Smith interviewed the Moon astronauts. He writes: 'When I review my travels among astronauts, my mind's eye goes first to the Houston shopping mall where Alan Bean sat for hours after returning from space, just eating ice cream and watching the people swirl around him, enraptured by the simple but miraculous fact that they were there and alive in that moment, and so was he'.

Note, however, the relative absence of official concern for the returning astronauts. The 'space medicine' offered by NASA has tended to lose interest in the astronaut's body once the job is done and the astronaut returns to normal life on Earth. Returning astronauts, especially in the early days of space exploration, have sometimes been seriously neglected, left to look after themselves once they re-entered society on Earth. This is another instance of the astronaut's body being objectified, attributed 'super' qualities, but at the same time, degraded. All these plans are instances of a project based on assumptions regarding humanity's supposed needs to be in the cosmos. These assumptions need intensive and critical discussion, to say the least.

Returning to Earth and Threats to Astronauts' Well-Being

To make these matters even more critical and urgent, conflicts between the controllers and the body in outer space do not end when the astronaut returns to Earth. In fact the process of re-entry may bring severe crises of its own, often deeply affecting the astronauts' mental as well as physical health.

Smith's 2005 study of the nine then surviving lunar astronauts is very suggestive. All the Moonwalkers considerably struggled to adjust to the social regimes and rhythms they faced when returning to Earth. This was partly due to the 'overview effect' of perceiving Earth and humanity from such a literal and figurative distance (White, 1998). These effects apparently often result in epiphanies experienced by many astronauts. For example, Marc Garneau, the first Canadian to enter outer space, reported: 'there are wars going on, there's pollution down there, but these are not visible from up above. It just looks like a very beautiful planet' (cited by White, 1998).

One way in which managers exert control over their astronauts is to make their experience seem more like what Johnson calls 'a microcosm of home life' should be made available to astronauts. Special occasions drawn from everyday life on Earth such as holidays and family birthdays will be shared between astronauts and their families. And selected football or baseball games will be beamed up to spacecraft via satellite for spacefarers to enjoy or otherwise. NASA managers actively encouraging these kinds of celebrations and practices, hoping that such events may in some way help to relieve monotony and sustain morale in the cosmos.

This links to the key issue of this chapter. Relations between Ground Control over astronauts can be very fraught, with the former having access to extensive knowledge and information which is not available to astronauts. So the issue of power remains, with the knowledge available to Mission Control bringing authority over their astronauts. Emancipation for astronauts will also remain unrealised so long as their bodily rhythms and practices are made subject to the 'appropriated time' and 'conflicting rhythms' imposed by Mission Control. Unrealistic assumptions about the bodies of astronauts being 'flexible' and capable of overcoming or endlessly adapting to the rhythms and dangers of outer space no longer suffice. Better relations, based on the experience of real human beings and their interactions with their spacecraft in the cosmos (and for increasingly long periods), have had to be pursued by mission control.

But none of this is to say that an autonomous astronaut, such as that represented in space fiction, is in sight. Furthermore, if the forces of capital are increasingly engaged in the process of relating bodies to the cosmos, this will surely support some relations and processes and not others. And astronauts also find ways to resist the controllers' authority.

Yet astronauts also find ways to resist the controllers' authority. In considering the relationship between power and human agency, Lefebvre introduced the notion of 'appropriated time', which has a special significance for the outer spatial body. It is, in Lefebvre's words, 'time that forgets time, during which time no longer counts (and is no longer counted)'. The reflections of Marc Garneau, quoted above, offer an example of 'appropriated time', in opposition to the highly organized 'linear' control over time directed by the managers from Earth. This is partly, of course, because it has evolved on Earth but has to undergo completely new demands and pressures when placed in the cosmos.

A range of critical questions arises when it comes to astronauts, particularly when they are removed from Earth's gravity and placed in outer space. What happens to an astronaut's body when it encounters the new and rapidly changing sequences of 'days' and 'nights' resulting from a spacecraft's changing relations with the Sun? There are some critically important political as well as purely 'scientific' issues arising. And they were exemplified by the cruel exposure of bodies to outer space during the Nazi regime.

The astronaut's body now receives a great deal of attention from not only space scientists but from Earth-bound controllers. The basic question here is how can long-distance space travel be made consistent with the body and its own internal rhythms? The two inevitably clash with the rhythms of day and night experienced by astronauts. But this puzzle raises further issues. A quarter of a century after the Nazi era, a 'science' of the human body in space has prevailed, particularly in the United States. Its long-term aim seems to be the establishment of a lasting human presence in the cosmos. But this raises major questions about what such a presence might be achieved for human beings. And, bearing in mind the increasing engagement of private capital into erstwhile 'government' practices, what are the limits to humans being transported into the cosmos?

And the above discussion relates to further questions. What is the future astronaut actually doing in the cosmos? The dominant assumption

6 Cosmic Capitalism and the Body

seems to be that he or she will be either undertaking new and exotic forms of tourism in the cosmos. Alternatively, the astronaut will be conducting 'scientific' work, including observations of the cosmos. But, as mentioned earlier in this study, the most likely possibility is that our astronaut will be undertaking military or 'defence' practices as discussed in Chap. 5. Symptomatically, NASA's Artemis project started life, for example, as another kind of Apollo project, one reaching the Moon for scientific purposes with wide-scale popular support. But this project is now being linked with *Space Force* and its military objectives. Furthermore, all these projects (governmental and 'civil') are, as discussed in Chap. 4, in large part the product of private capital.

And adding to these combinations and tendencies are straws in the wind, suggesting alternative future for humanity in the cosmos, one linking capital accumulation and 'defence'. Asteroid mining or capital invested in the extraction of rare materials from the planets may at some distant stage take place. But what started as a programme for accessing the cosmos for 'scientific' purposes is now being made into a different kind of project assisted and managed by the military–industrial–space complex. The social, political and environmental implications of all these developments are extremely serious. But they are difficult to imagine, especially when we are transfixed with developments in the far reaches of the cosmos while remaining uninterested in the closer-by use of the cosmos to enhance capital accumulation and use nearby satellites and military equipment to these ends.

The social and spatial relations of outer-space travel are thus being reshaped, as the balance of power between controllers and controlled has shifted in the latter's favour. The aim is to foster a more integrated connection between the astronaut's body and outer space. A deeper separation between the astronauts' lives in outer space and their earthly rhythms, on the other hand, has still not been fully bridged. But again, why is this necessarily a problem? It is part of a mind-set assuming an infinite progression into the cosmos. In the future, capital may or may not benefit from such a progression but that is surely a more than adequate reason to stop further adventures into the cosmos.

References

Anonymous Astronaut Quoted in Charles Fishman (2015, January/February). 5,200 Days in space. *Atlantic, 57*.

Association of Autonomous Astronauts. (2016). *Here Comes Everybody! The First Annual Report of the Association of Autonomous Astronauts*. Inner City AAA.

Barger, L., et al. (2014). Prevalence of Sleep Deficiency and Use of Hypnotic Drugs in Astronauts Before, During and After Spaceflight: An Observational Study. *Lancet Neurology, 13*(8), 904–912.

Casper, M. J., & Moore, L. J. (1995). Inscribing Bodies, Inscribing the Future: Gender, Sex and Reproduction in Outer Space. *Sociological Perspectives, 38*(2), 311–333.

Clement, G. (2011). *Fundamentals of Space Medicine*. Springer.

Dickens, P. (2015). Alternative Worlds in the Cosmos. In R. Vidal & I. Cornils (Eds.), *Alternative Worlds: Blue Skies Thinking Since 1900* (pp. 355–382). Emerald/Peter Lang.

Dickens, P. (2019). Social Relations, Space Travel, and the Body of the Astronaut (Chap. 9). In E. Cohen & S. Spector (Eds.), *Space Tourism. The Elusive Dream*. Emerald Publishing.

Dorrian, G., & Whittaker, I. (2018). Mars Mission: How Increasing Levels of Space Radiation May Halt Human Visitors. *The Conversation, 94052*.

Dunnett, O., & Maclaren, A. (2017). Towards an Anthropology of Space: Orientating Cosmological Futures. *Geographies of Outer Space Conference*, Department of Anthropology, UCL, London, 18.9.17.

Durrani, H. (2016). *Space Crystals and 'Our Window on the World'*. M. Phil Thesis, Department of History and Philosophy of Science, University of Cambridge.

Fishman, C. (2015). 5,200 Days in Space. *Atlantic Magazine*. http://www.theatlantic.com/magazine/archive/2015/01/5200-days-in-space/303510?

Flynn-Evans, E., et al. (2015). Circadian Misalignment Affects Sleep and Medication Use Before and During Spaceflight. *npj Microgravity, 2*.

Fong, K. (2013) *Extremes. Life, Death and the Limits of The Human Body*. London, Hodder and Stoughton.

Fong, K. (2014a). *Extreme Medicine: How Exploration Transformed Medicine in the Twentieth Century* (pp. 229–230). Penguin.

Fong, K. (2014b). The Strange Deadly Effect Mars Would Have on Your Body. *Wired Magazine*. https://wired.com/2014/02

Grady, M. (2017). Private Companies Are Launching a New Space Race – Here's What to Expect. https://phys.org/news/2017-10-private-companies-in-space.html
Hersch, M. (2012). *Inventing the American Astronaut*. Palgrave.
Hiltzik, M. (2015, December 28). 'The Day Three Astronauts Staged a Strike in Outer Space. *L.A. Times*.
Houston, R., & Heflin, M. (2015). *Go Flight! The Unsung Heroes of Mission Control*. University of Nebraska Press.
https://www.bbc.com/future/article/20171027-the-stray-dogs-that-paved-the-way-to-the-stars
Hu-Guo, J., et al. (2014). Keeping the Right Time in Space. *Military Medical Research, 1*(23).
Jacobsen, A. (2014). *Operation Paperclip: The Secret Intelligence Program that Brought Nazi Scientists to America*. Little, Brown.
Johnson, P. (2010, September–October). The Roles of NASA, Astronauts and Their Families in Long Duration Missions. *Acta Astronautica, 67*(5–6), 561–571.
Johnson, M. (2015). *Mission Control. Inventing the Groundwork of Spaceflight*. University Press of Florida.
Kevles, B. (2006). *Almost Heaven: The Story of Women in Space*. MIT Press.
Klinger, J. M. (2017). *Rare Earth Frontiers. From Terrestrial Subsoils to Lunar Landscapes*. Cornell.
Ladkin, A. (2011). Exploring Tourism Labour. *Annals of Tourism Research, 38*(3), 1135–1155.
Lefebvre, H. (1991, 1974). *The Production of Space*. Blackwell.
Lefebvre, H. (2004) (1991). *Rhythmanalysis: Space, Time and Everyday Life*. Elden and Moore, G. (Trans). London, Continuum.
Linnares, D. (2011). *The Astronaut. Cultural Mythology and Idealised Masculinity*. Cambridge Scholars.
Loomis, E. (2017). This Day in Labor History, December 28th 1973. www.lawyergunsmoneyblog.com/2015/12-this-day-in-history-december-28-1972
Lord, M. G. (2007). Are We a Spacefaring Species? Acknowledging Our Physical Fragility as a Step to Transcending It. In J. Dick & R. Launius (Eds.), *Societal Impact of Spaceflight*. National Aeronautics and Space Administration. Office of External Relations, History Division.
Marx, K. (1974). *Early Writings*. Penguin.
Marx, K. (1976). *Capital* (Vol. 1, p. 283). Penguin.

Moreno-Villanueva, M., Wong, M., Lu, T., Zhang, Y., & Honglu, W. (2017). Interplay of Space Radiation and Microgravity in DNA Damage and DNA Damage Response. *npj Microgravity, 3*, 14.

Mullane, K. (2006). *Riding Rockets*. Scribner.

Ormrod, J., & Dickens, P. (2019). Space Tourism, Capital and Identity (Chapter 10). In E. Cohen & S. Spector (Eds.), *Space Tourism. The Elusive Dream*. Emerald.

Peldszus, R., et al. (2014). The Perfect Boring Situation. *Acta Astronautica, 94*(1), 262–276.

Pesterfield, C. (2016). Cosmofeminism: Challenging 'Patriarchy in Outer Space'. In P. Dickens & J. S. Ormrod (Eds.), *The Palgrave Handbook of Society, Culture and Outer Space* (pp. 167–187). Palgrave Macmillan.

Phillips, R. (2011). *Grappling with Gravity: How Will Life Adapt to Living in Space?* Springer.

Phillips, R. (2012). *Grappling With Gravity*. Springer.

Rowen, R. (2017). *'Owning the Final Frontier?': 'The Emerging Geographies of Outer Space' Conference*, Department of Anthropology, UCL, London, 18.9.17.

Sage, M. (2014). *How Outer Space Made America*. Ashgate.

Smith, A. (2005). *Moondust: In Search of the Men Who Fell to Earth* (p. 347). Bloomsbury.

Stack Exchange. (2020). https://space.stackexchange.com/questions/20821/what-kind-of-time-regiment-schedule-do-iss-astronauts-have

Vaughan, D. (2016). *The Challenger Decision: Risky Technology, Culture and Deviance at NASA*. University of Chicago Press.

White, F. (1998). *The Overview Effect* (Vol. 230, 2nd ed.). American Institute of Aeronautics and Astronautics.

Whittaker, D. I (2018). Radiation Could Kill off a Manned Mission to Mars. *'T' newspaper*, 5.4.18.

7

Cosmic Risk Society

One way of understanding society's new relations with the cosmos is to adopt Ulrich Beck's influential concept of a 'risk society' (1992). Beck's account started with a critique of an approach to science undertaken in the seventeenth century by Francis Bacon, a distinguished English philosopher. He had a belief that science is essentially a progressive force, one which would blow away what he believed to be the mysticism and superstitions of his time. But Beck had a serious problem with such an optimistic view.

Facts and Truths

Bacon believed in science as establishing the 'facts' or 'truths' about nature and society. This would be achieved by detached observation of the world, a process which would evade fanciful guesswork and speculation about the social and natural worlds.

Scientific method, Bacon argued, would be a progressive force. It consisted of three main steps: first a description of 'facts'; second, a tabulation, or classification of facts into three categories—instances of the presence of the characteristic under investigation, instances of its absence, or instances of its presence in varying degrees, and third, the rejection of whatever appears in the light of these tables, not to be connected with the phenomenon under investigation. In these ways, a determination would be made of the causes of particular phenomena.

Ulrich Beck's Critique of Scientific Method

However, Bacon's emphasis on the exhaustive cataloguing of 'facts' provided no means of bringing an investigation to an end or pinpointing the causes of particular effects. More generally, Beck argued that the optimism originally associated with Francis Bacon's form of science has now dissolved.

The promises of a better and more peaceful world stemming from 'science' have not only failed to materialise. They have in many ways made matters, even worse, Beck would have argued, by the application of science itself causing an even greater array of crises and risks. And nowadays, Beck argued, unexpected and dire consequences are a regular feature of contemporary society, as scientists and governments alike do not adequately understand the risks they unintentionally cause. As a result, the authority of science as defined by Bacon has been thoroughly undermined.

Risk and the Chernobyl Disaster

The 1986 Chernobyl disaster was judged by Beck as an example of risk being *actually created* by science. It was a major nuclear accident that occurred in the Chernobyl Nuclear Power Plant, near the city of Pripyat in northern Ukraine, USSR. In fact its explosion is now considered to be the world's worst nuclear disaster in terms of casualties and costs.

Chernobyl was an example of modern society based on science almost literally blowing up in its face. And it was a result of modern society simply not having the scientific understanding—and hence the necessary control—over nature as it claims. Today's risks, Beck argued, stem less from natural dangers and hazards. They stem more from uncertainties generated by society's social development and over-dependence on 'science'.

So Beck argued that a disaster such as Chernobyl is a symptom of modern scientific understanding, one in which 'science' is not infallible and likely to produce catastrophic outcomes. Most important to the argument of the present study, these crises cannot be simply laid at the door of capitalism. Rather, Beck argued, they are a product of an over-dependence on 'science' and its promises of a better world. Furthermore, according to Beck, no one, including capitalist companies and government agencies, takes responsibility for these crises.

As a result, developing crises are not understood. But also in Beck's words,

> a fate of endangerment has arisen in modernity, a sort of counter-modernity, which transcends all our concepts of space, time, and social differentiation. What yesterday was still far away will be found today and in the future 'at the front door'.

All this is what Beck called 'organised irresponsibility'. Today's 'second modernity' society, he argued, is left coping with the unanticipated results and catastrophes of its earlier overly self-confident, 'scientific' enterprises and interventions and unforeseen outcomes. Enlightenment science promised great advances resulting from the implementation of science and technology. But, Beck argued, science now tends to be deployed as a kind of lifeboat to rescue society from the impacts of previous disastrous 'scientific' applications.

Capitalism and Risk

But there surely remains a major difficulty in Beck's argument. It fails to recognise the importance of capitalism in the generation of risk. It does not feature in Beck's work.

Power plants, for example, are there to provide energy for capitalist enterprises looking to make a profit. In other words, Beck's causal chain generating risk needs to be taken back to its sources. In the case of

Chernobyl, this reactor provided an energy source for what was then a socialist or communist society. Similarly, today's demand for ever-increasing energy and the dependence on nuclear power stems from the seemingly never-ending extension of capitalism, its cultures and its lifestyles.

In other words, the causal chains underlying risk needs to step well back from the immediate disaster and to investigate its underlying causes. These are nowadays the production of ever more commodities for sale. Beck talks of the 'manufactured', risk made in contemporary society. But it is surely the mass manufacture of commodities for sale to a market which is generating 'manufactured' risk. In short, it is capitalism and it seemingly unending production of commodities and endless amounts of 'stuff' which is generating risk. The result is global warming and the creation of large amounts of junk, often particularly afflicting developing countries.

Risk and the Cosmos

How do the above arguments stand up when we consider humanity's relations with the cosmos? Beck's 'risk society' concept can be extended to the cosmos but it remains problematic.

For example, a vast number of communications satellites have been launched by Elon Musk as a response to the supposed 'need' for citizens to communicate instantaneously around the world. These and other satellites for communication and surveillance have resulted in a huge amount of 'space junk', with satellites (many of them now disused) littering thousands and thousands of tons of space debris. These are not just disused communications satellites. They are also derelict spacecraft, spent rocket stages, exploded motors and other pieces of hardware jettisoned into the cosmos and forgotten totally. And they are now travelling in a circuit around Earth at speeds of 25,000 km per hour. Future spacecraft or orbiting satellites will, it is hoped, be able to capture these debris. But no practical measures to stop this kind of 'manufactured risk' being created.

But this raises a central critique of what Beck calls risk society. What has happened in the above case is that spacecraft have been made at a profit by private capital. And they have exploded in ways which are seriously dangerous to other spacecraft, satellites and of course to astronauts. So 'risk' in this case stems from capitalism itself and the imperative that it should at all costs survive and expand. But at the same time, owners of capital feel no responsibility to clear up the mess it has made in the cosmos. And 'solutions' have not yet been found because it has not yet been profitable to find them.

There is no obvious 'cause' as required by Francis Bacon. And until major disasters take place, and causing mass casualties, it seems the issue of manufactured risk is unlikely to be taken very seriously. It is most likely to be taken seriously as and when it is realised that there are profits to be made by clearing up the junk. At that point, cosmic capitalism may become interested in clearing up the debris it has left behind.

Perhaps the matter will start to be taken seriously when a serious accident in the 'nearby' cosmos actually happens and, with serious threats to human beings. In 2003, for example, the International Space Station was nearly destroyed by collisions with space junk. But despite these near-misses, so far representative of capital have taken no interest in clearing up the dangerous waste that cosmic capitalism has left behind. It is not a profitable enterprise to undertake.

Nuclear Risk

Matters could of course get completely out of hand when humanity extends ever further into the solar system. To this end, nuclear power is the most likely source of energy, this despite the recent development of safer propulsion methods based on solar energy. But the risks associated with nuclear power in the cosmos are not easily estimated and, once an accident has happened, it may well be very unclear who or what should be held responsible. And even if such a person or institution could be found what could be done about it?

There remain further risks associated with extension of society into the cosmos. In 1997, for example, a NASA Environmental Impact Statement was made in relation to the Cassini mission which investigated Saturn and its Moons. Its journey included a 'slingshot' manoeuvre which was accelerated by flying round Earth. In the interests of acquiring scientific data about Saturn and its moons, it was deliberately crashed on to Saturn's surface. Cassini was powered by Plutonium, a radioactive chemical element. Here the risk—including the risk of a plutonium-based accident—was not only recognised but still implemented.

In recent years, space scientists have been enlisted into various attempts to counteract some of the problems caused by science and industry and their effects of the cosmos. These problems, in keeping with the ethos of Beck's 'second modernity', are addressed by means of scientific and technological solutions but now on a much greater scale. One idea is that 16 trillion glass discs might be launched into orbit in order to deflect harmful rays from the Sun, which would otherwise reach Earth. This solution is, needless to say, incredibly expensive, technically very difficult to introduce. Which agency is likely to undertake such a project?

Furthermore, and at this point, we return to Beck's concept of 'risk society', it could necessitate yet more pollution in the manufacture and launch of the cosmic discs. Space technology has also been implicated in some of the wilder plans by the U.S. government to control the weather. One idea is that satellite solar power could be used to heat the air in a tornado to dissipate it. The U.S. Air Force is apparently also looking at schemes to control the weather for military purposes or 'warfighting applications'. But if Beck's theory of risk is in these cases correct, this will almost certainly have wider, supposedly unanticipated, even devastating, outcomes.

Beck's work clearly elucidates some of the contemporary issues stemming from space flight and space exploration. These particularly address the 'manufactured risk' stemming from space programmes. But his approach does not address the key role of capital and, since we are dealing here with a cosmic form of capitalism, an application of Beck's work is relatively limited in this case. This means we must again turn to political economy perspectives such as those introduced earlier in this study. These focus on the links between capital and governments and the implications for a socialised outer space. Beck's perspective does not fare well in this context.

The manufacturing chain and the whole of a person's life and her or his politics needs to be considered if we are to understand their politics and the political alignments involved. The politics endorsed and supported by rocket designers and rocket producers are an empirical question, not necessarily directly stemming from their work as scientists and rocket engineers. Their support for anti-war and anti-capitalist politics is certainly not guaranteed because such a project will threaten their jobs (Wills, 2016).

What about society's further extensions into the cosmos?

Space Junk as 'Manufactured Risk'

Even more important today is the growing crisis over space debris or 'space junk'. Here, Beck's prognosis of a risk undermining the promises of a first modernity seem at first to be coming true. In 1999, it was estimated that there were some 110,000 potentially damaging artificial objects hurtling through space. These included old spacecraft, rocket bodies and miscellaneous items left by early space missions and explosions generated by collisions in outer space. There are now many millions of pieces of such rubbish circling around the Earth in low orbit. The debris poses a substantial risk to people on the ground. An example is the hazardous material left on the ground by the space shuttle Columbia exploding across East Texas.

The United Nations has acknowledged the hazards of space debris, but so far little has been done to tackle the problem. The problem here is again capitalism. There's no money to be made in the investigation of hazards and the United Nations remains in effect. The 'best' solution for manufacturers and citizen seems to that of taking out some insurance against accidents and pass the bill on to shareholders.

The colonisation of space therefore mirrors the kind of risk society generated by colonisation on Earth. Activists, such as those in the Global Network Against Weapons and Nuclear Power in Space, are taking up the issue with a large section of their website devoted to the issue. The humanisation or conquest of space also well embodies Beck's notion of 'organised irresponsibility', the problem again being a result of an apparently highly scientific enterprise needing yet more scientific interventions to cope with the consequences.

Cosmic Capitalism and Health Risks

'Space junk' is also a substantial risk to satellites and space vehicles, especially the space shuttle. In 2003, the International Space Station was nearly destroyed by collisions with space junk. It was even argued at that time that the problem of surplus material circling round the Earth could place the whole space programmes at risk, a speculation which has led the European Space Agency to inaugurate a 'comprehensive solution' to the problem. Space debris mitigation procedures have been drawn up but are yet voluntary and not enforceable. The risks to astronauts' well-being are now quite well publicised. Less spectacular, but arguably more far-more reaching, are the risks generated by the current space research programmes.

Environmental activists, for example, campaigned against the environmental destruction caused by rocket launch emissions. In Russia, for example, children's illnesses have been linked to rocket launches. Children living near the Baikonur Cosmodrome in Kazakhstan are twice as likely to need medical attention as a result of high rates of hormonal problems and blood diseases. On launching, the rockets released hydrazine, a fuel said to be 'nasty and toxic'. A tablespoonful of this substance in a swimming pool is said to be capable of killing anyone drinking the water. Similarly, traces of rocket fuel chemicals have been found in milk and lettuce grown in Arizona and in bottled spring water from Texas and California. In Beck's terms, these are all instances of 'modern' innovations causing havoc and devastation. Companies might well say they had no idea that their chemicals would be so dangerous but this of course suggests that capital did not properly investigate the problems stemming from 'science'.

Capitalism and Risk

Beck's theory brings many insights. But, especially if we consider the expansion of capitalist society into the cosmos, we need to reconsider Beck's critiques. On this scale, there is in principle any number of ways in which a 'cosmic risk society' might be made. But in fact, since there are no, or very few, people entering the cosmos, there is frankly no serious need to act on Beck's theory. And this is a serious point. Why does it matter if the cosmos loses its pristine qualities? There still seems to be an underlying argument here that the cosmos should be kept in a pure and pristine state, one unsullied by humanity. But why should this necessarily be the case? Is it a hangover from the times when the outer reaches of the cosmos were deemed 'heavenly'. Or is it just pandering to a very ancient, quasi-religious, idea that the cosmos is supposed to be 'pure' and 'pristine' and must be kept as such.

There are some major issues for debate here. Capitalism is necessarily an expanding and crisis-making type of society. But capital, in order to continue reproducing and expanding, necessarily encounters limits. But in what sense does this really matter? It only really matters because the extension of society into space is nowadays the extension of capitalism into space. Marx wrote that 'the tendency to create the world market is directly given in the concept of capital itself'. He was saying that every limit appears as a barrier to be overcome.

If this is the case, the main reason for keeping the cosmos 'pure' is to stop the extension of capitalism. As discussed in Chap. 1, today's main power blocs (the United States, the European Union and in due course other societies such as China and India) are beginning to scramble for outer space in much the same way as the European societies competed for African territory in the eighteenth and nineteenth centuries. The establishment of property rights is central, as indeed it was when the African continent was subdivided by rival powers. Shammas and Holen argue that this process is now developing into the far distant cosmos. The question is that of 'deplanetising' capital based on Earth and spreading it indefinitely into the cosmos.

But the analysis of the present study suggests that is not really the main issue. There is actually no reason or evidence to suggest that capital is seriously looking to extend indefinitely into the cosmos. Rather, as detailed in other chapters of this study, it is 'piggybacking' or combining with projects (often military projects), with relatively 'nearby' projects which do not extend all that far beyond the Earth's surface. But perhaps this more 'localised' focus makes the issue even more difficult to resist than Shammas and Holen suggest. How to stop capital combining with the military is an urgent but even more difficult project than stopping capital entering the far-distant reaches of the cosmos. The military is now combining a range of political and economic interests. And this implies that 'conventional' political struggles on Earth will be combined with contestations (including military contestations) in the 'nearby' cosmos.

References and Further Reading

Adam, B., Beck, U., & Van Loon, J. (2000). *The Risk Society and Beyond. Critical Issues for Social Theory.* Sage.
Beck, U. (1992). *Risk Society: Towards a New Modernity.* Sage.
Marx, K. (1976, 1859). *Capital: A Critique of Political Economy.* Penguin.
Melman, S. (1970). *Pentagon Capitalism.* McGraw-Hill.
Shammas, J., & Holen, T. (2019). One Giant Leap for Capitalistkind: Private Enterprise in Outer Space. *Palgrave Communications.* https://doi.org/10.1057/s41599-019-0218-9
Sorensen, C. (2020). *Understanding the War Industry.* Clarity.
Wall, M. (2020). X-37B: The Air Force's Mysterious Space Plane. Space.com/ space.com/25275/-37bspaceplane.html
Wills, J. (2016). Satellites and Outer-Space Capitalism. In P. Dickens & J. Ormrod (Eds.), *The Palgrave Handbook of Society, Culture and Outer Space.* Palgrave.

8

Satellites, War and Capital Accumulation

The Military–Industrial–Space Complex

Previous and later chapters of the present study demonstrate in more detail the fact that 'nearby' parts of the cosmos are now being used for a wide range of new purposes. These particularly include military and civil communications of all kinds. They also include military personnel directing missiles on to 'enemies' literally thousands of miles away from the controllers. The whole process depends on military satellites.

As discussed in the previous chapter, major private corporations are nowadays a central part of the space industry, particularly the American space industry. And the American government presides over, and actively funds, this coalition between capital and military projects in the interest of 'defence'. These links between capital and outer space projects are a central part of capital's socialisation of the cosmos. But they are also a cosmic re-run of capital's wars and the making of empires, approximately a century ago.

History: Capital Accumulation and the Making of Empire

Some of the most insightful accounts of war and capitalism were those outlined by Lenin and Bukharin. Lenin's 1915 pamphlet, Imperialism: the Highest Stage of Capitalism and Bukharin's Imperialism and World Economy (1919) both argued that military competition between capitalist nations was rooted in the competition between nations for markets. Furthermore, then as now, governments and their armed forces were drawn in by capital as a key means of organising capital and labour in such a way as to engage in wars and, in the process, accumulate substantial profits.

Lenin and Bukharin showed that military competition between states for markets led to smaller capitalist enterprises being swallowed up and incorporated into much larger companies operating on an international scale. Lenin and Bukharin went on to theorise what was then the next stage of capitalism, one in which the expansion of capital into distant regions of the world needed the armies and weaponry which were required by capital to protect their assets. So, at this stage, the major imperialising governments such as Britain made armies and military equipment for further stages of capital accumulation.

The outcome was the global expansion of capital, with back-up from governments in the form of armies, weaponry and tax breaks for the private armaments industries. And as capitalist societies further expanded into new zones (including South Africa as part of the British Empire), small capitalist enterprises were swallowed up into large monopolies. At this stage processes of capital accumulation started to be intertwined with governments backing the interests of 'their' companies in distant regions.

In these ways, governments built up major alliances with 'private' or capitalist companies making new kinds of military equipment against possible—if often ill-undefined—'enemies'. Using the above kind of conceptual framework, Lenin argued that World War I (1914–1918) was 'an annexationist, predatory, plunderous war' between empires, whose historical and economic background must be examined 'to understand and appraise modern war and modern politics'. For capitalism to generate greater profits than the home market could yield, banks and industrial

8 Satellites, War and Capital Accumulation   107

cartels were merged to produce the so-called finance capitalism. This meant creating a large banking system in London with the central aim of exporting capital to countries with 'underdeveloped' economies. In colonising undeveloped countries, business and governments engaged in geopolitical conflict over the exploitation of labour throughout the populations across Africa and beyond. For Lenin, therefore, imperialism was the highest (or 'most advanced') stage of capitalism, requiring monopolies to exploit labour and natural resources and the export of finance capital.

The concept and the scale of the above theories need holding on to, if we are to understand what is happening now with an even 'higher' version of capitalism. Most obviously, today's spread of capital cosmos now needs incorporating into the kind of analysis developed by Lenin and Bukharin. A cosmic version of imperialism is now being made, with dominant societies, especially the USA, using satellites as a means of observing, attacking and controlling 'enemies' often thousands of miles away. Most importantly, note also that these satellites and the equipment involved in this new form of imperialism are made by capitalist companies in the USA, Britain and elsewhere.

It is obviously important to understand the specifics of what is now taking place, even though there are clear continuities with the analyses made by Lenin and Bukharin. On the one hand, capital continues to be introduced to parts of Earth in which materials are exploited. This means extracting resources on Earth which have not yet been opened up, either because they have been privately owned or because they are difficult to access. But the most obvious difference between today's form of imperialism and that outlined by Lenin and others is that no masses of people in the cosmos are waiting to be recruited and enslaved into capital's forms of labour processes into capital's forms of labour processes. Though movies such as Independence Day and Battlestar Galactica of course still assert the continuing presence of malevolent forces in the cosmos.

And here lies the most obvious contradiction within today's cosmic form of capitalism. Capital needs labour power if it is to survive, but in the cosmos, there is actually no labour power available for investment and exploitation. And if only a relatively few people are accessing the cosmos and its materials, it means that cosmic projects, such as mining and materials extraction, the Moon and the planets could look like distinctly unprofitable ventures. And the 'rare materials' in the cosmos would need to be extraordinarily rare and numerous to justify investment in large-scale cosmic forms of extraction. Furthermore, some kind of spacecraft would be needed to bring these precious materials back to Earth. These craft would be made in rocket factories on Earth, making the costs for capital even more considerable. All this militates against the classic form of imperialism suggested by Bukharin and Lenin and directly transposed to outer space. But the underlying relationships and processes prevail.

Nearby Outer Space: A New Zone of Capitalist Imperialism

The relationships between capital accumulation and war have long been a recurrent theme within critical social thought. As outlined above, war is best seen as an extension of capital accumulation, with the production of new kinds of weaponry being invented to fend off real or imagined threats made by real or imagined 'enemies'. But the 'classic' imperialism model advanced by Lenin and others needs to be considerably modified if it is to be made relevant for a cosmic form of capitalism. Furthermore, the hunt for 'rare' resources in the cosmos is not happening. A few exceptionally wealthy people may continue to circle Earth, or even visit the Moon and Mars, albeit at very great expense. Similarly, some claims may be made by companies or governments claiming ownership of the Moon, Mars, a nearby asteroid or in due course the Asteroid Belt.

But again, this is the old model of imperialism applied to capitalism and the cosmos. And none of this is happening on any scale now. And again, there are few if any signs that it will take place in the near future, not least because the costs of accessing the Moon, Mars or an asteroid would be, to coin a phrase, 'astronomic'. And even if extra-terrestrial resources were extracted, they would still need inputs into a labour process either in the cosmos or back on Earth. Capitalists would legitimately face some serious questions from their shareholders about such a plan.

The links between capital and the military remain exceptionally strong. And it is these links which demand an immediate and continued attention if we are to understand what is now taking place. Alternatives for a cosmic form of capitalism as spelt out above are best understood if we focus on what is already taking place. This, first of all, means recognising that surveillance, preparing for wars and engaging in wars, and using 'nearby' parts of the cosmos to these ends are already an optimal means of using the cosmos for capitalist ends. The satellites circling the Earth, of course made by capitalist companies, is an apt example of how capitalism relates to the cosmos today. They are now, thanks to the many forms of communications satellites circling Earth, a key means of capital accumulation. The same applies to satellite-guided 'Reaper' drones causing havoc and death amongst supposed 'enemies' on Earth. These drones are again made by capitalist conglomerates, and they have been very effective as a means of killing enemies in far distant regions without any risk of U.S. or British armies getting physically involved.

So for the foreseeable future, a human presence in the cosmos will also be used, particularly by the U. S. government and its allies, to assert military power. And the use of satellites by the U.S. military will be directed towards this end by a number of societies. This represents the form of cosmic capitalism today. The scene is now set for using the cosmos as a base for surveillance and warfare between dominant nations on Earth. And the alliances and beneficiaries will be much the same as those today. They are again big capital investing in the production of war and surveillance technologies, and they are paid for doing so by the U.S. government. Military personnel now work hand-in-hand with representatives of private capital to make new forms of military and surveillance equipment.

The Military–Industrial–Space Complex

Following World War II, President Eisenhower famously described 'The Military Industrial Complex' as a deeply problematic force in American society. He was referring to the increasingly close integration between the U.S. government and the very large privately owned corporations making and selling military equipment. Eisenhower had regularly encountered this 'complex' while serving in prominent military positions during the war.

In 1970, Seymour Melman published *Pentagon Capitalism*, a study which largely concurred with Eisenhower's earlier analyses. 'In the name of defence', he argued, 'and without announcement or debate, a basic alteration has been effected in the governing institutions of the United States. An industrial management has been installed in the federal government, under the Secretary of Defence, to control the nation's largest network of industrial enterprises'.

Melman was describing the long-term and very close relations between government and the U.S. military. It was one in which the forces of capital and state are combined in the interests of fending off and eliminating supposed 'enemies', all in the interest of 'peace'.

The September 11 attacks on Manhattan of course further legitimated such merging. It was clear, in that case, that there were real enemies of the United States out there, even though individuals other than Osama Bin Laden were difficult to identify and kill.

Furthermore, and as Sorenssen argues for the present day, the forces of capital have now largely 'captured' the group of senior advisers who are determining the future of outer space. The integration between capital is now immensely powerful. As Sorenssen puts it,

> Corporate executives and operatives at the National Space Council and its Users' Advisory Group (UAG) worked behind the scenes to shape policy. The former is comprised mostly of Cabinet Secretaries and other high-ranking government officials whose departments and agencies have been largely captured by corporate interests, War Industry executives powered by the UAG. (2020: 352)

Sorensen goes on to describe the USA's Space Development Agency, an organisation which combines the needs of the military at The Pentagon with the profit-making requirements of private capital. 'The 2019 Missile Defence Review Agency', Sorensen writes, 'which hyped up the traditional array of "threats"—Russia & China and Iran & North Korea, the "great powers" and "rogue states", respectively—in order to justify "strengthening" and "expanding" missile defence, including spending an ungodly amount of money on, for example, layers of low orbit satellites. While mentioning war corporations' products like the SM-6, the writers of the Missile Defence Review were careful to not mention war corporations' names. Corporate personnel operating inside the Office of the

Secretary of War in effect helped craft the U.S. Missile Defence Review'. The new U.S. Space Force was also subjected to review. As Sorensen puts it 'at no point is the need for and development of Space Force an issue which is free of corporate demand for profit and the force of its influence' (Sorensen, 2020, op.cit: 353). So this is a worked example of the prime relationships between society and the cosmos today. Capital is not being invested in outer space itself. Rather, it is latched on to a military presence in outer space as means of making profits. This is the prime means by which capital invests in the cosmos today.

Furthermore, and behind the extensive military manoeuvring in support of the space capitalism, Sorenson also shows in greater detail, the devastating fact that the private sector in the U.S. space industry nowadays dominates what used to be government-led enterprises.

So what is 'good' and 'safe' for American citizens has been conflated with what is 'good' and 'safe' for U.S. corporate capital. In the case of the space sector, this link between capital and government is now being achieved, courtesy of the National Space Council and a 'Users Advisory Group' which together recommends which military innovations should be made and used to 'keep America safe'. Sorensen clearly supplies the answers. Fusions between the military and private capital are now absolutely central to 'keeping the peace'. In this way, 'peace' and capital accumulation are very closely intertwined but, as Sorensen insists, capital accumulation is very much the main and dominant force in this liaison between government and capital.

Boeing, Lockheed Martin, Northrop Grumman and *Raytheon* are what Sorensen (2020) calls 'The Big Four' as regards designing, developing and making new space-based technologies. Boeing makes and runs military communications, satellites as well as civil and military aircraft of many kinds. Lockheed makes and sells satellites for communications purposes and also for tracking other satellites and monitoring so-called space junk. It also has what Sorensen calls 'deep space-based relations' with elements other states including the South Korean *Defence Intelligence Command* and Britain's *Defence Intelligence*. *Northrop Grumman* makes satellites for monitoring weather conditions and missile launches. And *Raytheon* 'prides itself on having a foot in all parts of the space programme'. (Sorensen, 2020: 347). It focuses on selling electronics and software systems underpinning space systems. It also makes radar, satellite communications and navigations systems. 'All four of the aforementioned corporations work on lucrative, highly classified satellite projects, which are kept out of the public domain' (Sorensen, 2020: 347). The U.S. *War Department* 'purchases worldwide satellite communications services from such corporations as Arte LLC, SES Government Solutions, Imarsat and Iridium'. Military satellites are launched from the Evolved Expendable Launch Vehicle, a joint venture between Lockheed Martin and Boeing. For reasons that remain unclear, it has only once been launched: once in its whole life.

So the dominant and now familiar argument here is that military equipment and spacecraft paid by private capital and government are once more being used to 'keep America 'safe'. It is believed by government that no country or organisation would dare to attack the United States because they know fully well that American rockets and hardware would simply wipe them out. And it is fair to say there is considerable popular support in the United States for armaments production. Not only does it keep America safe but, as Sorensen's work shows and as described by Bukharin and Lenin a century ago, wars and preparation for wars greatly enhance capital accumulation within the so-called private sector.

Meanwhile, popular support for such space programmes is clear from a recent Pew Report. This demonstrates a considerable enthusiasm not only for the private space industry but also for its extending links with U. S. government. As the report puts it, 'the majority of Americans believe it is essential that the U.S. remains a Global Leader in Space'. Furthermore, the increasing role of private companies in space exploration is not seen as in any way a major problem. U.S. investors are happy to invest in the new, privately owned companies making rocketry. They believe these investments are made to keep the people 'safe'.

There nevertheless remains a coherent and persuasive alternative argument. This was made by the Campaign for Nuclear Disarmament in the 1960s and 1970s, and it is still being made now. They make a completely contrary argument, one insisting that increased investments in warfare are more likely to create rather than stop war. For CND, the most effective and long-lasting form of peace would be one which has abolished the weaponry used in war. Furthermore, as further discussed in the final chapter of this study, it would convert the skills of the armaments manufacturers into making products which are socially progressive rather than militaristic.

All these arguments and developments require further revealing and debate, not least because there remains much mystery surrounding the links between private capital and the military. Why is the space industry so attractive to the American people and its governments? Why do they hold to the idea that a presence in outer space is necessarily 'a good thing'? Or, as suggested by Sorensen and by his arguments cited above, is there an extra underlying argument involved here? The American electorate seems to seriously believe that not only will armaments in the cosmos keep America safe but the capital accumulation in the space industry will again help keep their lives and investments 'safe'.

In such ways, government and capital, combined with widespread backing from the American government public, create and maintain a cosmic form of capitalism, one that appears to be very successful in economic, governmental and popular terms. Everyone, or so it seems, is 'safe' as a result of enhanced levels of armaments production.

Such, then, are the social and political contexts and coalitions in which the United States' so-called the war industry operates. The U.S. government regularly incorporates corporate interests in its decision-making. And the corporations, by design, continue to strive for maximum profit. So again, the profitable nature of war is what propels the plans made by the military–industrial–complex, on the one hand, and the U.S. congressional complex, on the other.

The military–industrial–congressional complex (MIC) is a powerful but little-known group. It consists of the U.S. military establishment (headquartered in the Pentagon); private industry (what Soren calls 'the war industry'), corporations that market and sell goods and services to the U.S. military and the intelligence agencies. Capitol Hill and its elected representatives fund the military and pass legislation, and they are now the central part of a seemingly permanent 'warfare state'.

But for all the rhetoric surrounding money and 'safety', there remain some positives to be recognised and celebrated. The production of landers due to alight on Mars and elsewhere requires the skills of the machinists working at The Jet Propulsion Laboratory, California, as described in

Chap. 5. More widely, it requires some organisations to be relatively well insulated from capitalism if scientific missions are to succeed in accessing the cosmos. Again, the most relevant example is the skilled machinists working at The Jet Propulsion Laboratory, as also discussed in Chap. 5 and in the conclusion of this study.

New and progressive kinds of relations between capital, labour and governments in the space industry are being made, but these are primarily in the European industry, one which includes workers making spacecraft as a part of corporatist 'deal' between capital, labour and government. But such a deal is very unlikely in the case of the U. S. space industry. Again, the closest, and now most quickly growing, relations involved spacecraft production are those between representatives of the U.S. government and representatives of capital. This is the coalition outlined throughout this chapter: between big capital, on the one hand, and government military and surveillance projects, on the other—all in a supposed 'national interest', that of arming and protecting the U.S. public against aggressive but as yet unknown 'enemies'.

Conclusion: Lessons from the Mysterious Space Plane

The Boeing X-37 'Space Plane', operated by U.S. Space Force, is one of the latest and more spectacular projects based on the abovementioned alliance between financial, industrial and military interests. It gets special coverage because this plane is operating in a somewhat curious way in 'nearby' parts of the cosmos.

The plane is launched into the cosmos aboard a large spacecraft designed and built by the ubiquitous private corporation, United Launch Alliance as discussed in Chap. 5. Unlike satellites, the X-37's orbit takes an elliptical form, this making the craft less predictable for an enemy hoping to destroy this spacecraft. After completing its mission, the Space Plane returns through the atmosphere, landing on a runway like a conventional aircraft.

But what is this craft really about? Heather Wilson, a former U.S. Air Force secretary, has tried to lighten the tone by helpfully informing reporters that the plane's elliptical circuits are 'designed to make enemies of the United States military to go crazy. Our adversaries don't know where it's going to come up next. And we know that drives them nuts. And I'm really glad about that' (https://uk.finance.yahoo.com/news/heres-x-37b-spaceplane-disappears).

But apart from driving America's enemies go crazy, what else does the new Space Plane do? According to one insider, its objectives are actually somewhat mundane. It is testing out new kinds of military equipment. But this account still does not enlighten us as to what the new Space Plane does and will eventually do. The project received the enthusiastic backing of President Trump and of the private company discussed in the previous chapter, United Launch Alliance.

Few people seem to know what exactly The Space Plane will be doing. But perhaps that is the whole point. What this and similar 'defence' projects do very well is (1) swell the funds of a private rocket company such as United Launch Alliance, (2) encourage private investment into 'government' projects to 'keep America safe' and (3) invest in satellite-guided drones designed to survey and if necessary eliminate distant enemies. In short, this plan is a modern version of that described by Lenin and Bukharin. In both cases, the power of capital and government is being combined to threaten and even engage in war.

Meanwhile, Britain now has its own version of keeping everyone 'safe', based on the U.S. model. In April 2021, the British military has supplemented the USA's Space Force with its own 'Space Command'. Made up of the three existing military forces plus a representative of Space Command, its facilities have recently been located at Fylingdales, Yorkshire, a picturesque part of the countryside previously used to monitor missile launches from The Soviet Union. Now it is recruited for a different kind of emergency, that of tracking any 'enemies' threatening the peace and using a presence in the cosmos to this end. As I write, it may be examining satellites and manoeuvres connected to President Putin's invasion of Ukraine.

References and Further Reading

Balibar, E. (2010, March/April). Marxism and War. *Radical Philosophy, 160*.
Bukharin, N. (1915, 1929). *Imperialism and the World Economy*. Monthly Review Press.
Caldicott, H., & Eisendrath, C. (2007). *War in Heaven. The Arms Race in Outer Space*. New Press.
Campbell, S. (2014). Why Does Capitalism Lead to War? *Socialist Worker*, Issue 394.
Dawson, L. (2018). *War in Space. The Science and Technology Behind Our Next Theatre of Conflict*. Springer.
Geppert, A. C. T., Brandau, D., & Siebeneichner, T. (2021). *Militarizing Outer Space. Astroculture, Dystopia and the Cold War*. Palgrave.
Johnson-Frees, J. (2017). *Space Warfare in the 21st Century*. Routledge.
Ledbetter, J. (2011). *Unwarranted Influence. Dwight D. Eisenhower and the Military Industrial Complex*. Yale University Press.
Lenin, V. I. (1915, 1973). *Imperialism. The Highest Stage of Imperialism*. Penguin Great Ideas.
Marx, K. (1976, 1859). *Capital: A Critique of Political Economy*. Penguin.
Melman, S. (1970). *Pentagon Capitalism*. McGraw-Hill.
Pew Report. (2018). The Majority of Americans Believe It Is Essential that the U.S. Remains a Global Leader in Space.
Reno, J. (2019). *Military Waste. The Unexpected Consequences of Permanent War Readiness*. University of California.
Shammas, J., & Holen, T. (2019). One Giant Leap for Capitalistkind: Private Enterprise in Outer Space. *Palgrave Communications*. https://doi.org/10.1057/s41599-019-0218-9
Sorensen, C. (2020). *Understanding the War Industry*. Clarity.
Wall, M. (2020). X-37B: the Air Force's Mysterious Space Plane. Space.com/ space.com/25275/-37bspaceplane.html
Webb, D. (2009). Space Weapons: Dream, Nightmare or Reality? In N. Bormann & M. Sheehan (Eds.), *Securing Outer Space*. Routledge.
Wills, J. (2016). Satellites and Outer-Space Capitalism. In P. Dickens & J. Ormrod (Eds.), *The Palgrave Handbook of Society, Culture and Outer Space*. Palgrave.
Wirbell, L. (2004). *Star Wars. U.S. Tools of Space Supremacy*. Pluto.

9

Cosmic Capitalism and Space Law

Earlier chapters have outlined the social and political relations of today's cosmic capitalism. As we saw in Chap. 3, accessing the cosmos and using its riches to enhance human well-being back on Earth is a familiar trope. It continues today but it defies logic because it is by no means clear that Earth has actually 'run out of resources'. And it is not even clear that the cosmos contains the resources needed by capitalists.

Yet logic does not always prevail when decisions are made about society's relations with the cosmos. This chapter is important because The United Nations made a farsighted ban on the commercialisation and militarisation of outer space. This ban continues but it is increasing pressure.

The Space Treaty in Historical Context

Space law is a body of law governing space-related activities. Like international law more generally, it comprises a variety of international agreements, treaties and conventions within the United Nations General Assembly.

The term 'space law' refers to the rules, principles and standards of international law appearing in the five international treaties and five sets of principles governing outer space. These have been developed under the auspices of the United Nations as follows.

The U.N. Outer Space Treaty

The most important of these treatises and principles are contained in The United Nations 1967 *Outer Space Treaty*. This Treaty still stands as a bedrock to society's relations with the cosmos. In this way, space law addresses a variety of matters such as the preservation of the space environment and liability for damages caused by space objects, the settlement of disputes, the rescue of astronauts, the sharing of information about potential dangers in outer space and international cooperation. Here is part of the UN statement explaining the overall scope of this Treaty.

> Treaty on Principles Governing the Activities of States in the Exploration and Use of Outer Space, including the Moon and Other Celestial Bodies.

The Outer Space Treaty provides the basic framework on international space law. It is based on the following principles, as set out by the Treaty:

- the exploration and use of outer space shall be carried out for the benefit and in the interests of all countries and shall be the province of all mankind.
- outer space shall be free for exploration and use by all States.
- outer space is not subject to national appropriation by claim of sovereignty, by means of use or occupation, or by any other means.
- States shall not place nuclear weapons or other weapons of mass destruction in orbit or on celestial bodies or station them in outer space in any other manner.
- the Moon and other celestial bodies shall be used exclusively for peaceful purposes.
- astronauts shall be regarded as the envoys of mankind.

- States shall be responsible for national space activities whether carried out by governmental or non-governmental entities.
- States shall be liable for damage caused by their spacecraft or satellites.
- States shall avoid harmful contamination of space and celestial bodies.

The Space Treaty's fundamental and highly ambitious principles have since guided the whole conduct of space activities. This includes, as stated earlier, space envisaged as 'the province of all humankind' with space exploration and the use of outer space by all states 'without discrimination'. Also included is 'the principle of non-appropriation' of outer space by nation states. Some of these principles and relationships have been seriously tested but the 1967 Outer Space Treaty as outlined above remains the key piece of international legislation affecting society's relations with the cosmos.

Talks on preserving outer space for peaceful purposes began in the late 1950s at the United Nations. The U.N., on behalf of its member nations, submitted proposals in 1957 to reserve outer space for exclusively 'peaceful and scientific purposes'. Table 9.1 is the terms of the Treaty.

Table 9.1 The United Nations Treaty on Principles governing the activities of states in the exploration and use of outer space, including the moon and other celestial bodies

The Outer Space Treaty provides the basic framework on international space law. It contained the following important principles:
  the exploration and use of outer space shall be carried out for the benefit and in the interests of all countries and shall be the province of all mankind.
  outer space shall be free for exploration and use by all States;
  outer space is not subject to national appropriation by claim of sovereignty, by means of use or occupation, or by any other means.
  States shall not place nuclear weapons or other weapons of mass destruction in orbit or on celestial bodies or station them in outer space in any other manner.
  the Moon and other celestial bodies shall be used exclusively for peaceful purposes.
  astronauts shall be regarded as the envoys of mankind.
  States shall be responsible for national space activities whether carried out by governmental or non-governmental entities
  States shall be liable for damage caused by their space objects.
  States shall avoid harmful contamination of space and celestial bodies.

The Soviet Union initially rejected the Treaty because it was at the time preparing to launch its first satellites and, more importantly, to test its first intercontinental ballistic missiles. But after a period of negotiation between the United States and the Soviet Union, governments agreed a form of words which allowed the Treaty to be signed.

So as early as the mid 1950s, support for the U.N. *Outer Space Treaty* stemmed not only from questions of science or even prestige. Crucially, the UN Treaty also facilitated the so-called commerce as part of its objectives. And, at the same time, it also resisted the use of the cosmos for military purposes.

But these aims and objectives did not come from nowhere. This UN legislation should be seen in the wider context of corporatist forms of state intervention which were important at the time.

Corporatism and the Cosmos

The 1960s and 1970s was the era of 'corporatism', a form of politics in which national governments brought representatives of capital and labour together in a 'trilateral' form of decision-making. Together, the representatives hammered out the future of their society. Centrally important was the collaboration between capital and labour in a supposed 'national interest' (Jessop, 1990).

So the intention of governments including the British government subscribed to this kind of collaboration. This was made possible if everyone, including Trade Unions and company owners, collaborated around a compromise between capital and labour. The compromise centred on agreed levels of capital investment and a workforce which would desist from industrial action, if levels of investment and pay were agreed. And all this was in the context of a shared 'national interest', one which would supposedly benefit everyone.

But there were variations in the type of collaboration. The outcomes were dependent on the balance of class forces in different nations. In Scandinavia, for example, the unions were strong and the construction companies were highly industrialised and productive. This element, combined with exceptionally high levels government, provided extensive welfare provision (Dickens et al., 1985). But while such corporatist collaboration was important at the time, such arrangements were often weakened and even removed, depending on the balance of class relations in different societies. But corporatist arrangements, while successful for a time, were always a difficult compromise to keep, especially for unionised labour which remained deeply suspicious about collaboration with capital. But it remained in place, especially in the Scandinavian societies.

Though elements of corporatism still exist today (that of the Scandinavian societies is a good example), corporatism has been watered down and virtually killed off in many societies from the 1980s onwards, with monetarist politics being used to privatise public assets and, with a weakening of Trade Union power, exercising control over labour forces.

Corporatism and 'Success' of the Cosmos

But the key point here is that corporatist politics continued to prevail when it came to outer space policy. Prior to the official introduction of The Space Treaty, The United States and Soviet governments submitted separate drafts of outer space treaties to the UN General Assembly in June 1966. A mutually agreed 'treaty-text' was worked out over a period of six months, with the UN General Assembly giving its approval of the Treaty on December 19, 1966. The Treaty was signed in Washington, Moscow, and London on January 27, 1967, and it entered into force in October 10 of that year.

So in corporatist terms, access to outer space was a 'success' in terms of developing new knowledge of and a degree of control over the humanisation of the cosmos. Any interventions would, it was assumed, continue to benefit society as a whole. And until quite recently, this has remained a relatively intact and shared ambition to which all UN member-states could sign up. And this focus has to a large degree remained in the present day.

In these ways, a strong sense of collectivity, idealism and shared interests was incorporated into the United Nations Space Treaty. And to a large extent, these shared values have been maintained in the Treaty today. But they are now being potentially undermined by the growing use of outer space for military purposes, as outlined in Chap. 6 where the militarisation of the United States' Space Force was discussed. But while the UN legislation forbids the establishment of military bases and 'manoeuvres' in outer space, it does not expressly ban all military activities in space itself. In this sense, the developments, such as US *Space Force*, appear to be in the clear.

The Outer Space Treaty

In 1963, the UN General Assembly approved two resolutions regarding outer space. These subsequently became the basis for the Outer Space Treaty. UN Resolution 1966 called on all countries to refrain from stationing Weapons of Mass Destruction in outer space. The UN Resolution set out legal principles on outer space exploration. It stipulated that all countries have the right to freely explore and use space.

A central feature of this UN resolution was *terra nulius*, or 'land belonging to no-one'. So Article II asserted that 'outer space would not be subject to national appropriation by claim of sovereignty, by means of use or occupation, or by any other means' (Goswami & Garretson, 2020, op.cit. p. 72). In short, outer space would still be considered as *common property*. This was part of the UN's corporatist 'deal' regarding nations' differing relations to outer space.

In more detail, The Space Treaty's main aim was to control provisions in the United Nation's Article IV. This demands that states must not:

- place in orbit around the Earth or other celestial bodies any nuclear weapons or objects carrying Weapons of Mass Destruction (WMD).
- install WMD on celestial bodies or station WMD in outer space.
- Establish military bases or installations, test 'any type of weapons' or conduct military exercises on the moon and other celestial bodies.

Other treaty provisions further emphasised that the cosmos should not be a single country's domain. All societies have a right to explore the cosmos. These provisions stated that:

- Space should be accessible to all countries and can be freely and scientifically investigated.
- Space and 'celestial bodies' are exempt from national claims of ownership.
- Countries must avoid contaminating and harming space or celestial bodies.
- Countries exploring space are responsible and liable for any damage their activities may cause.
- Space exploration should be guided by 'principles of cooperation and mutual assistance', such as obliging astronauts to provide aid to one another if needed.

In these ways, societies were assembled around a shared, corporatist, notion of the cosmos as used for peaceful and collaborative purposes. At the time, there was little if any serious question of the cosmos being used for military and, more particularly, commercial purposes.

Like other UN treaties, the Outer Space Treaty allows for amendments or member withdrawal. Article XV also permits countries to propose amendments. But an amendment can only enter into force if accepted by a majority of states parties, and it will only be binding on those countries that approve the amendment. (In this way, the original corporatist ideal of consensus regarding occupation of the cosmos remains.) Article XVI states a country's withdrawal from the Treaty will take effect a year after it was submitted with a written notification of its intentions to the 'depositary states': the United States, Russia and the United Kingdom.

Article IX of the UN Resolution asserted that the exploration and use of outer space, including the moon and other celestial bodies should be 'guided by the principle of co-operation and mutual assistance and shall conduct all their activities in outer space, including the Moon and other celestial bodies with due regard to the corresponding interests of all other States Parties to the Treaty'. In this way, the corporatist nature of space politics remains clear. And the language of 'co-operation and mutual assistance' remains to this day.

The Treaty came into force on October 10, 1967. Today it has 110 'states-parties', or countries, that have ratified or acceded to the Treaty. Through ratification or accession, a country agrees to be legally bound by the Treaty's provisions. Another eighty-nine countries have currently signed up in principle to the Treaty's ideals.

The Treaty bans the stationing of weapons of mass destruction (WMD) in outer space. And it also prohibits military activities on celestial bodies. Furthermore, it also focuses on binding rules which govern 'peaceful exploration' and use of space. In this way, the original corporatist and collaborative ideals of The Space Treaty have been largely retained.

The Treaty also forbids countries from deploying 'nuclear weapons or any other kinds of weapons of mass destruction' in outer space. The term 'weapons of mass destruction' is not defined but it presumably refers to nuclear, chemical and biological weapons. The Treaty does not explicitly prohibit the launching of ballistic missiles through space, but it repeatedly emphasises that space is to be used for 'peaceful' purposes. This has led some analysts to conclude that the Treaty can be interpreted as prohibiting all types of weapons systems, not just WMD, in outer space. Significantly for our concerns with outer space, however, the 1967 UN legislation did not explicitly interfere with nations extracting the resources of the cosmos.

Such, then, were the corporatist ideals transferred to the regulation of society's links to the cosmos. Note the continuing emphasis on collaboration and cooperation, this again reflecting the UN's original corporatist ideals. These corporatist shared ideals remain. And they still require that any extraction of resources should benefit *all* of the Earth's inhabitants. For the moment, the ideals of the 1960s prevail, and they remain a lesson for Earth-bound societies. But it now seems likely that they will be questioned and, almost 'literally', undermined.

The 1967 Outer Space Treaty: Recent Assessments

There are now in practice two positions as regards the future of the 1967 *Outer Space Treaty*. One, published around the start of the Trump presidency, remains very pessimistic about its prospects for keeping outer space as a 'commons'. Indeed Shammas and Holen (2019: 3) argue that the Outer Space Treaty (OST) and its promise of the cosmos as a 'commons' is now under very serious threat. The cosmos, they argue, is now being opened up for profitable ventures not by mankind but by what they call 'capitalistkind', with the demands of capital over-riding the notion of a shared and collective forms of ownership. So the status of the OST, Shammas and Holen argue, 'has been denied'. They argue that 'communism in space was a possibility only so long as space was materially inaccessible to capitalistkind'. But now, with space being opened up and potentially made into a site of profitable exploration and adventures, 'outer space proto-communism must falter and fade away' (2019: 3). So this suggests that the expansion of society into space is a straightforward and direct result of a developing cosmic capitalism, with investors of capital fuelling further expansion.

But a somewhat less gloom-ridden appraisal of the 1967 *Outer Space Treaty* has been made by Dunstan (2020). He points to the 1967 U.N. Outer Space Treaty (OST) as 'the cornerstone of space law'. And Dunstan takes a more optimistic view. 'Contrary to what some commentators claim', he argues, 'there is no "loophole" that allows individuals or capitalist organisations to claim ownership of celestial bodies'.

This is because Article II only addresses nations, not companies or individuals as potential owners. Dunstan also makes it clear that individuals themselves cannot make claims to celestial bodies, this because nations must both 'authorise' and 'supervise' the activities of its national organisations to ensure compliance with the provisions of the Outer Space Treaty.

These examples, combined with what Dunstan calls 'a proper reading' of the space treaties, demonstrate that domestic legislation such as that passed by the United States is entirely consistent with international space law and practice. He argues that there is actually no 'loophole' allowing individuals to claim ownership of celestial bodies. This is because Article II of the OST is concerned only with 'nations' and not with individuals. In fact, Dunstan argues, Article II specifically precludes citizens from making property claims in outer space. He also suggests that outer space is analogous to the high seas. It is 'a place where many travelled but no one owns'.

A Serious Corrective

So Dunstan offers what seems at first a very persuasive argument. But it is very important to step back and relate his argument (and indeed that of Shammas and Holen) to the supposedly 'rare' resources on Earth and the supposed 'need' to access the resources of the cosmos.

These writers are correct, in the sense that 'ownership' of outer space is not currently allowed, even now when there are landers trundling around and inspecting the surface of the moon and Mars to examine what these entities are made of. But, very important to this chapter's argument, neither authors take sufficient account of the 'rare resources' and the reasons for their rarity. To explain this further, we must turn back to the Earth and the supposed 'rarity' of its resources.

Earth-Bound Resources, Fictions and the Cosmos

To adequately understand the difficulties with Shammas' and Holen's argument and that of Dunstan, we need to seriously address Julie Klinger's pioneering work on what she calls 'the fictions' of rare materials on Earth (2017). The last chapter of her study of 'rare earth frontiers' starts with the observation that,

> no-one is actively mining the Moon. However at least six national space programmes, fifty private firms and one graduate engineering programme, are intent on figuring out how to do so. The fictions of rare earth scarcity emerging from the 2010 crisis gave lunar mining proponents something specific around which to frame their cause and to prospect for "Rare Earths" site. (Klinger, 2017: 199)

But crucially for her argument, Klinger goes on to describe these 'fictions' and their consequences. Most importantly, she suggests that Earth is actually *not* running out of resources. Furthermore, lunar mining proponents are 'trafficking in fictions, fears and erroneous assertions made in the face of established facts concerning both international legal conventions and the global rare earth supply'. Meanwhile, lunar mining proponents have made what Klinger calls 'significant progress…in moving the question of off-Earth mining from the fringes into the mainstream of political discourse and public consciousness across the globe' (2017: 199).

So again, Klinger argues that there are in fact plentiful 'resources' below the surface of Earth. But a crucial fact remains. The underlying 'problem' is the fact that these resources, particularly copper and lead, are present but often very difficult to find and extremely expensive to extract on Earth. This is due to a range of factors, not least the fact that Earth's 'rare resources' are often located in extremely inaccessible and difficult-to-mine places such as the Amazon area and Afghanistan. And just as crucial to Klinger's argument, these areas are either privately owned or sometimes difficult to access.

So all these mean that the 'crisis of rare materials' is to a degree a figment of the imagination. Again, 'rare materials' on Earth are not absolutely rare. They are 'rare' either because they are in private ownership and/or they would be too costly for capital to gain access and start extraction. So again, this 'rarity' is socially made and not absolute. Furthermore, any 'rarity' is largely a product of private ownership and/or capital's reluctance to invest in areas difficult to access. And this means that drilling for rare materials on the moon or elsewhere is not 'essential' in an absolute sense, although it may be taking place. In sum, costs plus private ownership on Earth are actively *making* the 'rarity' of raw materials on Earth.

As regards the moon, Klinger points to 'The Lunar Frontier', *The Mare Procellarum KREEP* region which is supposedly 'high in rare earth elements, thorium and iron' (2017: 203). She also points to a range of similar lunar sites which are also being assessed by U.S. and Chinese government landers (See Fig. 9.1). It seems that not all these sites are being investigated solely for 'rare materials'. (Some may be occupied by governments for 'defence' purposes.) But ownership and drilling in the cosmos are again both premised on a supposed absence of rare materials on Earth. Again, they are a magical 'solution' to a 'problem' ('limited resources' on Earth) that actually does not really 'exist'. More accurately, it exists but it is too troublesome and costly for capital's profits being used to excavate and make into a half decent profit for the space entrepreneurs in the process.

'Off-Earth mining' used to be on the fringes of political discourse but now it has for some time become 'respectable'. This started when a

EXTRAGLOBAL EXTRACTION   203

EARTH'S MOON, NEAR SIDE

Mare Procellarum
KREEP boundaries

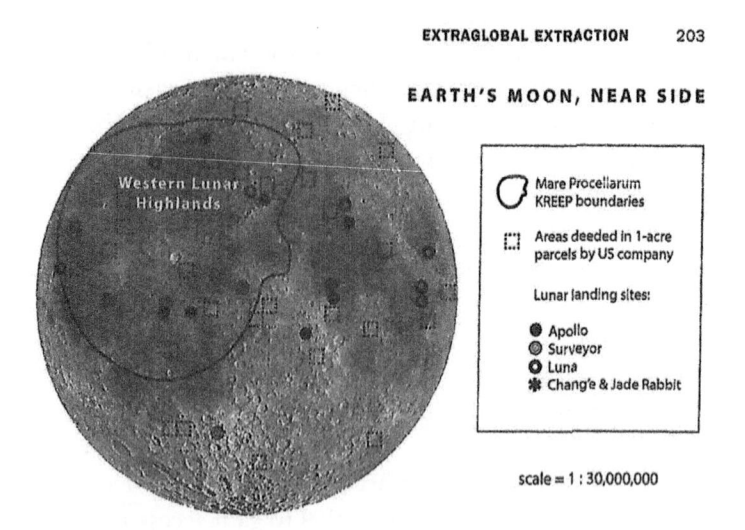

Areas deeded in 1-acre parcels by US company

Lunar landing sites:
● Apollo
◉ Surveyor
○ Luna
✱ Chang'e & Jade Rabbit

scale = 1 : 30,000,000

Fig. 9.1 Mining the near side of the moon. (Source: Klinger, 2017: 203)

U.S. company, Planet Resources, went public with a plan to mine asteroids. It was backed by billionaires, including some of those at the upper reaches of Google, the world's richest company at the time. In 2015, President Obama signed into law, a U.S citizen Competitiveness Act which allowed the commercial recovery of asteroid 'resources'. It was passed by the International Institute of Space Law, which argued that the Act was 'a possible interpretation of the Outer Space Treaty'. And soon afterwards, the International Institute of Space Law cleared it as a legitimate expansion of the original Outer Space Treaty. Soon afterwards, Luxemburg created its own law and an investment fund of $227 billion to lure space companies into its clutches. The U. S. and Chinese governments meanwhile stood by ready to undertake very costly lunar excavations. In the end, the plan was never put into practice, largely it seems because the original crisis to the U.S. economy had faded away. But it would come as no surprise if the 'rare resources on Earth' argument is brought back into play.

But there is no *absolute* need for such a 'solution' to the 'problem' of supposedly 'dwindling' resources on Earth. But such inconvenient facts are not important when it comes to understanding a cosmic form of capitalism. The 'problem' is confined to getting out into the cosmos and opening up yet more opportunities for capital accumulation. This despite the fact that such 'rare' materials were being left underground on Earth. But logic and rationality will probably never be capital's strongest point.

Klinger closely links the interest in lunar sites to the 2010 financial crisis (investment in these regions being seen as an implausible solution to this crisis) but again she is sceptical as to the real and imagined extent of lunar materials. She seriously doubts, for example, the assertion made by China Academy Professor Ouyang Ziyuan that 'The Moon is full of resources—mainly rare earth elements, titanium and uranium, which the Earth is really short of, and these resources can be used without limitation' (Klinger, 2017: 204).

And against these arguments, Klinger again insists, there is actually no absolute limit to 'rare earth materials'. The limit is again self-inflicted, a direct product of capital's ownership of Earth-bound resources and the risk of inadequate profits being made by such an undertaking.

So again, capitalist society may well be expanding into the cosmos on a false prospectus, but this does not of course mean that such expansion will not take place. Again there remains much mystification and wishful thinking about 'rare materials' on Earth when such 'rarity' is socially constructed and separated from the awkward facts of capital's ownership patterns on Earth. But this is not to deny that mining outer space remains a more exotic and glamorous pastime than grovelling around inhospitable places and awkward property rights back on Earth.

And for this reason, the attraction of 'rare materials' in space may well take place, even if it is entirely illogical and untrue. Klinger points out that, 'startup space mining companies are collecting billions in investment and government technology-transfer contracts' (2017: 203). All this with a view to further 'fixing' and accessing capital on the moon and in due course extracting its supposedly 'rare lunar materials'. Klinger puts the capitalists' dilemma well. 'Earth is not actually short of rare earths or platinum group metals, nor can lunar resources be used "without limitation" as though outer space were the ultimate no-consequences terrain of extractivist freedom' (Klinger, 2017, op.cit: 204).

Cosmic Capitalism: Nation States as Facilitators

Outer space law is different from law applied in nation states. Again, The Outer Space Treaty of 1967 is, like other space treaties, based on countries rather than relations between capital and labour. The United Nations in this way prevents countries asserting territorial rights over parts of space or a planet.

But now something new is happening, one appropriate for the new era when it is capital rather than nation states proposing to own parts of the cosmos to this end. Nation states are now setting themselves up as new facilitators of today's cosmic capitalism.

Dunstan (2020) points, for example, to The Duchy of Luxemburg's $227 million fund to attract corporations considering investments in parts of the cosmos. The Duchy's own account of this process is as follows.

> Luxembourg established itself very early on as a state fostering entrepreneurship and the development of commercial activities in the space sector. By supporting the creation of SES, one of the biggest satellite operators in the world, and by creating legislation specific to the transmission of satellite services shortly thereafter, Luxembourg has demonstrated its ability to build a favourable environment for the structural development of activities related to the use of outer space. (Dunstan, 2020)

A similar plan comes from the Dubai government. Dubai is a relatively small nation state investing in outer space projects. It is creating its own legal 'court' to assert its commercial space rights. They are part of Goswami and Garretson's so-called *Scramble for the Skies*. And these same authors and their co-contributors document 'the Great Power Competition to Control the Resources of Outer Space'. But Goswami and Garretson are getting a little ahead of themselves. The actual reality of resource-extraction companies mining the asteroids and the moon and making vast profits in the process remains to be seen.

First, and bearing in mind the work of Klinger, are there really resources out there on a scale attractive to capital? Second, and perhaps most importantly, would companies based in, for example, the United States and China, idly stand if the profits made by accessing the moon or the asteroid belt went to Luxemburg or to Dubai? The financial institutions of the United States and Europe are very well-established, and they would almost certainly take steps to stop the rewards for extracting the cosmos passing them by and going to Dubai and Luxemburg.

Yet, though the above work in Dubai and Luxemburg is assertive, it misses a much more important point, one which has informed this whole study. The 'competition' is now not so much between countries. It is now much more about capitalist organisations competing for the resources of the cosmos and using or making legislation towards this end. The Dubai government and its banking system is just a relatively small nation state investing in outer space projects.

Goswami and Garretson (2020: 1) inform their readers that 'a new space (is) just beginning—though one of vastly different character. It is not, as before, a race for prestige and honour among nations, or a contrast between ideologies. It is rather a race to secure the determinants of economic and military power between states' (2020: 1–2). In a superficial sense, it may look as though nation states are scrambling for the resources of outer space. But this again ignores the harsh realities of political economy. Any 'scrambling' will be done not by governments alone but, more importantly, by capitalist enterprises. Governments in this context are likely to have limited and facilitating roles.

The whole process of accessing the cosmos is in practice likely to part of an unpredictable interplay of interests. Kirgis (1996) explains how international law is now being enforced, with countries 'behaving badly' being policed by the UN with possible imposition of sanctions. But often, the U.N. Security Council is hamstrung by other issues that dominate relations between superpowers. At the time of writing, for example, President Biden is trying to stop Russia annexing Ukraine. But if 'China' joins with him in that process, he might have to turn a blind eye to Chinese extracting activities on the moon and elsewhere. Such trade-offs are a regular feature of United Nations and its laws, even in 'peaceful' times.

Conclusion

All this entails setting U.N. space law in the wider context of political economy. As Pashukanis (2001) points out, it is also useful to remember that law 'has an important but secondary role'. Law often makes claims to be neutral and operating in a 'universal' way. But claims of its 'universalism' can be misleading. Truisms about 'Earth's dwindling resources' and contrasts with 'the infinite abundance of the cosmos' offers little understanding of the extraction and labour processes taking place. These have been discussed in previous chapters. Similarly, space law may seem to be 'universal' but once we start examining how space law actually operates (particularly in relation to the larger countries grouped together in the United Nations and with ready access to these countries' to 'raw materials'), we may find that law's claims to be 'universal' start evaporating into thin air. The UN Space Treaty skates a little too rapidly over political economy and remaining overly focussed on states, or what Anderson (1983) called the 'imagined communities' of nations.

Capital's demands, and not those of governments, remain the principle driving force accessing the cosmos. But as Pashukanis (1978), Arthur (1978) and Chandler (2017) point out, it is also useful to remember that 'law' itself has a distinctive status. It is one which claims to be neutral and operating in a 'universal' way, not unlike the status given to commodities and money. But, as outlined above, these different forms of 'neutrality' and 'universalism' are suspect. Laws, especially space laws, claim to be operating, in general, literally in 'universal' ways when in practice they may be enhancing the well-being of the already powerful. Once we start examining the empirical detail of how space law is actually used, how it operates and who gains and who benefits in the process, we may find law's claims of universalism starting to unravel.

References and Further Reading

Baars, G. (2018). *The Corporation, Law and Capitalism*. Haymarket Books.
Balibar, E. (2010). Marxism and war. *Radical Philosophy*. https://www.radicalphilosophy.com

Bastani, A. (2020). *Fully Automated Luxury Communism.* Verso.
Collis, C. (2012). 'The Geostationay Orbit. A Critical Legal Geography of Space's Most Valuable Reall Estate (Chapter 3). In S. Parks & J. Schwoch (Eds.), *Down to Earth.* Rutgers.
Collis, C. (2016). *Res Communis? A Critical Geography of Outer Space.* Palgrave.
Dickens, P., Duncan, S., Goodwin, M., & Gray, F. (1985). *Housing, States and Localities.* Methuen.
Dunstan, J. E. (2020) 'Mining outer space may be cool but is it legal?' *Room: Space Journal of Asgardia.* Room.eu.com/article/mining-outer-space-may-be-cool but-is-it-legal?
Goswami, N., & Garretson, P. (2020). *Scramble for the Skies.* Lanham.
Jessop, B. (1999). *State Theory. Putting Capitalist States in Their Place.* Cambridge, Polity.
Klinger, J. M. (2017). *Rare Earth Frontiers. From Terrestrial Subsoils to Lunar Landscapes.* Cornell.
Lenin, V. I. (1973). *The State and Revolution.* Foreign Languages Press.
Pashukanis, E. (1978). *The General Theory of Law and Marxism.* Selected Writings.
Petersen, L. (2021). *The Future of Governance in Space.* New Degree Press.
Shammas, B., & Holen, T. (2019). One Giant Leap for Capitalistkind: Private Enterprise in Outer Space. *Palgrave Communications.* https://doi.org/10.1057/s41599-019-0218-9

10

Future Work: Cosmic Capitalism, Indigenous Peoples and Satellite Television

A cosmic form of capitalism includes capital's use of satellites and linked communications equipment to create satellite television. This chapter is a test of television communications. It examines whether, if at all, indigenous peoples (with their particular histories, lives and values) can relate to satellite television.

Satellite TV must make profits, and this means that indigenous peoples' needs are not catered for by capital in the form of television programming. So satellite TV directly conflicts with indigenous peoples.

Indigenous peoples' lives and cultures are oriented around particular regions, albeit vast regions. And uprooting them means detaching them from the settings in which they and their ancestors have lived. They are thoroughly uprooted and alienated from their histories, cultures and ways of life.

Society, Nature and Aboriginal Peoples

Aboriginal societies make no distinction between 'society' and 'nature' (see Fig. 10.1). There is just one world and it consists of other human beings, non-human entities such as animals and inanimate objects such

Aboriginal societies

```
              Other humans
                   |
             ( Human
               being )
            /         \
    Inanimate         Non-human
    entities          entities
  (rocks, water, air) (animals, plants)
```

Capitalist societies

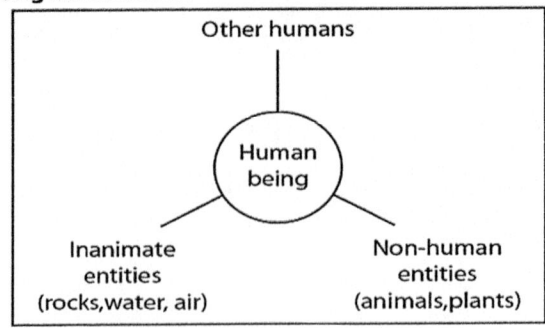

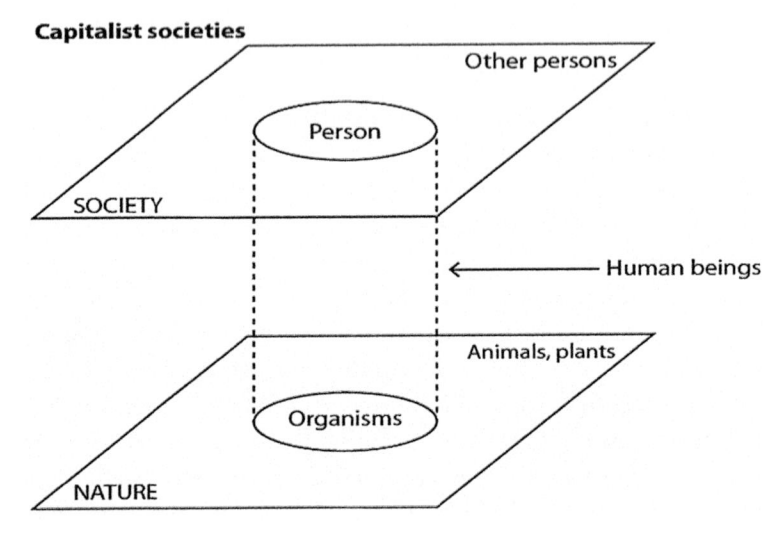

Diagram showing two planes: upper plane labeled SOCIETY containing "Person" and "Other persons"; lower plane labeled NATURE containing "Organisms" and "Animals, plants"; arrow labeled "Human beings" connecting between the planes.

Fig. 10.1 Society, nature and the body in indigenous and in capitalist societies

as rocks and streams. But in modern, capitalist societies, all the above characteristics have changed. First, there is an infamous split between 'society' and 'nature'. This split of course stems from capitalist societies 'using' nature in the interests of capital accumulation. (And what Marx called the 'revenges' of nature on society stemming from such separation are now becoming disastrously too apparent.)

10 Future Work: Cosmic Capitalism, Indigenous Peoples... 137

As Fig. 10.1 Aboriginal Societies suggests, 'people in modern capitalist societies' are considered part 'social' and part 'natural' and knowledge is represented in books and in theories. Knowledge is gained by *stepping out* of nature and making theories about it; this rather than learning from experience, interacting with the natural world and developing a more complete kind of 'knowledge' as a result. These differences are profound, and they are spelt out in Fig. 10.1.

Today, there are 370 million indigenous people around the world, spreading across more than 90 countries. Many experience harsh lives. These include separation from their ancestral lands, denial of opportunities to express their cultures and exclusion from the labour market. Indigenous peoples across the world are often treated as second-class, even in some cases 'subhuman'. This is a direct result of them being torn from their ways of live, as portrayed in Fig. 10.1.

Indigenous Peoples in Australia

These relations between society and the cosmos in indigenous societies were first spelt out in the early years of the twentieth century by the sociologist Emile Durkheim. He never went to Australia himself, but he studied and theorised the 'primitive' (nowadays called 'first') peoples in Australia. They lived, and many still live, in remote areas detached from 'mainstream' society.

And Durkheim argued that, the cosmic order and the social order were—and still are—considered to be one and the same thing. As we have seen in Fig. 10.1, they do not make a distinction between 'heaven' or 'cosmos', on the one hand, and an 'earth', on the other. Similarly, they do not make the 'modern' type of distinction between 'society', on the one hand, and 'nature', on the other. For indigenous peoples, they are one and the same. And for the indigenous peoples of Australia, the sky, the stars and the planets had special meanings for ancestors and also to modern indigenous peoples. The differences between societies of indigenous peoples and those of people in capitalist societies are stunning. As described in Fig. 10.1, aboriginal societies are organised around a world composed of other human beings, animals, plants and inanimate entities such as rocks and water.

The contrast between this and the situation in capitalist societies is very considerable. First, capitalist societies are conceptually divided into two levels, those of 'society' and those of 'nature'. The first is composed of human beings and the second is the sphere of 'nature', with the former dominating the latter. But note that the human subject is construed as part 'social' and part 'natural'. (A division of course reflected by the division between academic disciplines with, on the one hand, sociology and politics and, on the other hand, biology and physics and the sciences more generally.)

Understanding the world in the case of capitalist societies entails stepping out of the environment and learning about it through books and 'science' and perhaps even television rather than, as in the case of indigenous peoples, learning about it as a result of direct engagement and experience. The effects of separating the human being from the natural world are now all too apparent, with, for example, species loss and climate being a direct result 'man' and her or his change be the result to this separation. Marx and Engels were amongst the very few social scientists to argue that, on the one hand, humans are a 'natural' sort and their impact on the natural world is a result of treating it as a mere 'object' to be extracted and incorporated into labour processes. On the other hand, they are a 'social' sort, with their lives consisting of their many interactions with other people.

Indigenous societies such as those described by Durkheim and those encountered shortly in this chapter have close, interlocking, relations between the workings of their society and the structure and workings of the universe. But capitalist modernity systematically disconnects these peoples from one another and from their environments. Television exacerbates this profound division, picturing the split on television screens.

The result for humanity in a capitalist society has long been a direct alienation of internal nature (that of human beings) from external nature. Persons are simultaneously 'social' as well as 'natural'. But as Fig. 10.1 suggests, the world human beings live in is characterised by a radical division between the 'society' and 'nature'. The result is again the distinctive split experience for human beings in a capitalist society. They, unlike indigenous peoples, are no longer considered to be a part of nature. Their lives, forms of 'ownership' and consumption patterns separate them from the environment on which they depend.

Indigenous Peoples and Satellite Television

Some indigenous peoples in Australia work as labourers on cattle ranches used by their aboriginal ancestors. But other Australian aboriginal peoples live in settlements in small towns, often under poor living and working conditions. Again, the picture is one of alienation not only from their ancestral lands but also from the collective rituals which form a part of their culture.

So indigenous societies such as those described by Durkheim also have close, interlocking, relations between the structure and workings of society and the organisation of the universe. But again, these links are shattered by contemporary society which is based on the dichotomies of society versus nature as described above.

Satellite TV has for some time had a particular significant importance for indigenous peoples, this including the lives of 'First Peoples' now living in remote parts of the globe and regions of society including Australia. And there are some important social and sociological issues here. Indigenous or 'aboriginal' peoples are often relatively powerless and only largely separated from mainstream society. But at the same time, television seem to offer some, at least a partial, understanding of the cosmos.

'Indigenous' satellite television in Australia, as elsewhere, has reflected these peoples' marginal status and their separation from the natural world. But television stations have to work within the limits imposed by capitalism. To survive, 'Indigenous TV' stations have long been dependent on capital. But the financial and social circumstances of indigenous peoples' means they have not been able to attract the kinds advertising and programming available for commercial TV station. The result has been that indigenous television has a largely 'symbolic' value. At best, it offers a divided vision. In practice, capitalist modernity systematically disconnects these peoples from one another and from their ancient lands. The picture is therefore one of separation and alienation, one which satellite television can never mend.

Imparja Television: From Indigenous to 'Remote' Broadcasting

An Australian television station named Imparja (or 'footprints') owned and organised by indigenous peoples has encountered a host of problems, and these all arise from the separation between these peoples and their lands. Based in Alice Springs, at the epicentre of the Australian continent, Imparja TV has long been directed towards first peoples, their histories, priorities and values. It operates satellite programming, and it has for some time been a government-financed station, one organised and run by indigenous peoples.

Imparja TV has long been a symbolic centre for aboriginal Australians. It is a not-for-profit enterprise, and on this basis, it continues to transmit a small amount of material aimed at Aboriginal peoples. But Imparja's funds have been very limited, and this makes it necessary to include large sections of mainstream culture.

Table 10.1 A sample of Imparja TV schedule in July/August 1999

Time	Origin	Normal Day	Naidoc Week
4.00 pm	Imparja	Yamba	Yamba
4.30 pm	Aus	Pig's Breakfast	Pig's Breakfast
5.00 pm	Aus	Catchphrase	Catchphrase
5.30 pm	Aus	Neighbours	Neighbours
6.00 pm	Imparja	Imparja National News	Naidoc week
6.30 pm	U.S.	A Current Affair	Naidoc week
7.00 pm	Aus	Sale of the Century	Sale of Century
7.30 pm	U.S.	Veronica's Closet	
9.27 pm	Imparja	News/Weather	Imparja News
9.30 pm	U.S.		Dawson's Creek
10.30 pm	Aus.	Good News Week	
11.30 pm	U.S.	Melrose Place	Melrose Pl
12.30 pm	Aus	Sports Tonight	Sports
12.58 pm	Imparja	On Track	On Track
1.00 a.m.		Station Close	

Source: Parks (2005)
Yamba at 4 pm is a honey ant who is energetic and eager to learn. He has a close following amongst young TV Australian aboriginal people watchers

As Table 10.1 shows, Imparja's programming was also combined with a great deal of commercial material from non-indigenous sources, again because it cannot afford broadcasting put out by commercial television. So Imparja has long been contradictory. It has a degree of symbolic value for indigenous peoples, but at the same time, its programming largely disconnects them from their lives and their forms of knowledge.

Imparja TV is owned and administered by Aboriginal people and it is a not-for-profit enterprise promoting the well-being of its First Peoples. Parks 2005 argues that Imparja, created and run by indigenous peoples working at Imparja, has long been regarded as a symbol of Aboriginal Australia. But now, its history has recently been troubled, largely because of its lack of funds and investment.

Intensive restructuring has taken place, making Imparja into a so-called Remote television. Under a 'Regional Equalisation Scheme', the government minister was attempting to 'level up' Imparja-style programming to other channels. In this way, it was believed that financial position of indigenous peoples' broadcasting could be achieved. The minister introducing this process told Imparja that 'no matter where you are in Australia you should have access to the same free-to-air options'.

It was in effect demanding that the Imparja TV model should be extended over the whole of Australia over a much larger TV network. Such a project was tempting for Imparja because it would bring in more revenue.

But for Imparja peoples, such a project completely undermined its whole purpose. It meant spreading their meagre finances and selling their material still further to five channels, all of which would be purchased by a station with the name of 9 Network.

In these ways, the Australian government attempted to make Imparja TV into a 'universal' channel, operating on as large scale as possible. But again, whereas capital requires as 'universal' a kind of TV programming as possible, this was definitely not the case for Imparja. They were, and still are, a 'regional' station but now with a remit for a 'local' (and very large) audience of indigenous peoples. The original ideals and output of Imparja TV were shaky enough but now they were in practice made absent and irrelevant.

From Indigenous to 'Remote' Television: Alienation Continued

But the government's 'Regional Equalisation Scheme' was disastrous for the original ideals of Imparja TV. Again, 'equalisation' meant in practice broadcasting from not one but from five separate channels. This was obviously a very long way from Imparja's original project. Furthermore, all the information collected had to be sent to Sydney's Nine Network for processing and redistribution by satellite. The result was that Imparja had to serve an extra 500,000 people by extending from one to five channels.

So the original model for Imparja, one based on a relatively small-scale connection with its viewers, was now destroyed. Alister Feehan, the boss of Imparja and himself an indigenous person, said 'he struggled to make sense of what has been directed from up high'. As he put it, 'Truly local news and content have disappeared from Imparja's offerings yet we're in a place where stories are currency and the living lore goes back further than anywhere else in the world'. 'We don't even see ourselves as being Regional', Feehan explains. 'We now call ourselves Remote. How do you actually produce a news service? You don't. You can't make it relevant. So what we do is take Nine's News service, pull it to pieces and insert some local content'.

'We experimented two years ago', Feehan explains. 'We thought we would go down a user-generated news service where we would use local people in local communities to generate the stories. We had a couple of journos (sic) basically to check the validity of what we had, and produce a news service that way. We tried that with (Public Interest News Gathering funding) but we couldn't get funding for it from an industry point of view', *Alice Spring News* 29.10.20.

So the alienation resulting from the Imparja case was that social and cultural 'equality' for indigenous peoples was conflated with wholly vacuous idea of 'remoteness' or 'remote television'. Imparja now covers a landmass area of 3.6 million square kilometres east of the Western Australia border. At the time of writing, it has screens across six Australian states. Feehan summarised the position as follows:

So the market's 'solution' for Imparja TV on the one hand destroyed its original focus on indigenous life but at the same time to revivify it with so-called local content. But 'local content' is for sure not what indigenous peoples are looking for. An understanding of their histories and lives cover very large areas of central Australia are what is needed. Similarly, 'remoteness' is of no particular interest to indigenous peoples. It was if anything a disincentive for indigenous peoples to switch on to this channel at all.

But the owners of Imparja TV nevertheless made 'remoteness' as their key means of staying financially solvent, even though it thoroughly fails to meet indigenous peoples' needs. 'Imparja' now, for example runs Darwin's DTV Network as a 'local' service to cover the Northern Territory. But this area now covers 1.42 million square miles. It is a zone which means little to anybody, including the indigenous peoples of Australian Northern Territory. In the end, Imparja was inserted with a name and a very small input into much larger and more conventional TV stations whose aim was straightforwardly to increase market share and profits. The marginalisation of indigenous peoples' TV was thus made complete.

And that was surely a sorry way for a progressive and far-sighted project such as Imparja to finish up. The upshot has been distinctly poor for Imparja and its supporters, because its local and regional identity around indigenous TV were entirely lost in the whole process of reconstruction.

The new arrangement was summed up even more critically by Alice Spring News as follows:

ALICE SPRING NEWS
Imparja: once a relay station, the other broke. (By E. Chlanda)

Local television started in Alice Springs with a noble agenda: made-in-Alice news, current affairs and entertainment as well as Aboriginal cultural programming that would be towering over the inevitable influx of imported material, not measure in airtime but as what would make the first Aboriginal station very special.

Now 35 years later, Imparja is little more than a relay for a major network and its parent, the Central Australia Media Association is running a hapless fire sale to pay for its chronic debts.

In the early 80s Freda Blynn, a spokesperson for Imparja TV told the Broadcasting Tribunal, in a nutshell, that foreign trash should be kept out of the pristine lands of Ancient Aboriginal tradition which should get a boost from new story-telling medium in a region where stories are currency.

National Indigenous Television (NITV)

But in 2007, the position for indigenous peoples in Australia changed again when National Indigenous Television (NITV) was created. Note first, however, that the creation and funding of this channel was not simply a magnanimous government gift to the indigenous peoples of Australia. It was in fact a direct result of government concern about the exceptionally high rates of imprisonment of indigenous peoples. The assumption was that the new NITV would distract any future trouble makers and miscreants into a *national* system and values.

NITV includes the half-hourly nightly NITV News, with programming that includes current affairs programmes, sports coverage, entertainment for children and adults, films and documentaries. The stated objective of NITV is to bridge a perceived 'gap' between indigenous and non-indigenous peoples of Australia. But this is surely problematic. We are back here to the previously mentioned problem of 'remote' television. Again the differences between indigenous and non-indigenous peoples is not simply those of scale. Rather, it is one of power and direct connection with the natural world. And it remains very unclear whether a national television for indigenous people can adequately address the underlying relations between indigenous and non-indigenous peoples. First, it is important to understand whether and how indigenous peoples are adopting 'mainstream' cultures and ways of life. Yet a recent study by Bennett et al. (2021) suggests that more indigenous peoples actually are now watching 'mainstream' television. If this is correct, it must mean that their ancient connections with the natural world and their ancestors have been destroyed and transformed and in some sense incorporated into mainstream values and ways of life. But where does this leave the histories and concerns of indigenous peoples?

It seems the absence of the original version of Imparja (which, as we have seen, was itself originally combined with extensive commercial television output) would not be missed by the most recent generation of indigenous peoples. But it now seems that quite large numbers are watching government-funded Australian television in the form of NITV. Does this mean they are in some sense 'signed up' to Australian modernity? And is this a new generation of indigenous people who have abandoned their ancient connections? A similar point could be made about the numbers of non-indigenous peoples now apparently watching NITV.

But a focus on television in any form still does not adequately reflect the social needs and practices of indigenous peoples themselves. These are made clearer when we look at the much wider international picture.

Indigenous Peoples and Communications: A Continuing Struggle

Meanwhile indigenous peoples in the United States, Canada, the United States, Latin America and elsewhere have not been standing still. In fact, they have for some time been taking matters in their own hands. And not relying at all on new legislation or new TV channels.

Using hand-held recorders, video machines and the internet, they are making films, documentaries and animations to record and share their connections with the ancient lands. And now indigenous peoples, including those in Australia, are sharing their experiences and histories with other indigenous peoples across the globe. This is now a new kind of social movement opening up, one which is now operating on a global scale while at the same time recognising and supporting the diversity of indigenous peoples' histories and practices. Wilson and Stewart (2008) describe these new, and very positive, developments which involve an array of technologies.

Whether discussing Maori cinema in New Zealand or activist community radio in Colombia, native peoples use both traditional and new media to combat discrimination, advocate for resources and right, and preserve their cultures, languages and aesthetic traditions.

In short, indigenous peoples are now re-making their own lives *beyond* conventional programming such as that offered by Australia's NITV. 'First peoples' are again drawing on and sharing their ancient origins and practices described by Durkheim more than a century ago. So many are now increasingly now taking their pasts and their futures in their own hands rather than remaining dependent of capitalist televisions organisations offering so-called remote television or indigenous programming.

So here is a radical alternative to the tokenism of Imparja, 'remote broadcasting' and NITV. Indigenous peoples are using simple, hand-held cameras to describe and share their lives beyond capitalist video cameras broadcasting. They are using media in many forms featuring film, documentary, animation, video art, radio broadcasting and the internet. And indigenous peoples are tending to share their lives both within their localities while productively connecting with many other indigenous peoples around the world. In these ways, they are escaping from the clutches of conventional capitalist television broadcasting and its limited offerings of 'remote' television. But not only are they escaping from the likes of Australian NITV. They are at the same time relating to other indigenous peoples with the aid of relatively cheaper equipment.

So to an increasing extent, indigenous peoples are now taking the form of international projects in which they are organising, sharing their own lives and resisting 'official' government narratives. And most importantly for the wider topic of this study, they are escaping the clutches of global corporate capitalism and its limited offerings of satellite television.

On the one hand, indigenous peoples have long been denied satellite television in ways adequately related to their practices, their lives and their histories. But they are now finding new and valuable ways of sharing and enhancing their interests, values and histories. But others are now more fully engaging with a new, fully blown Australian television station which aims to incorporate 'indigenous peoples' cultures into mainstream modernity. This is an ongoing contest between the culture offered by satellite television and the very longstanding cultures of indigenous peoples. The end-result is as yet unclear but the power of indigenous peoples' ways of life against mainstream television is undoubtedly on the rise.

References and Further Reading

Baudrillard, J. (1983). *Simulations*. Semiotexte.
Bennett, T., Carter, D., Gayo, M., Kelly, M., & Noble, G. (2021). *Fields, Capitals, Habitus. Australian Culture, Inequalities and Social Divisions*. Routledge.
Dickens, P. (2004) Society and Nature. Changing Our Environment. Changing Ourselves. .
Dickens, P., & Parry, J. (1998). Ecological Competence in a Modern Age. A Role for the New Information Technologies. In A. Bolder (Ed.), *Work and Education Handbook 98 Ecological Competencies*. B & A Publications.
Gramsci, A. (1971). *Selections from the Prison Notebooks*. Lawrence and Wishart.
Miramax Video. Rabbit Proof Fence. DVD.
Regan, T. (2020). *Australian Television Culture*. Routledge.
Virilio, P. (2012). *The Great Acceleration*. Polity.
Williams, R. (2003). *Television: Technology and Cultural Form*. Routledge.
Wilson, P., & Stewart, M. (Eds.). (2008). *Global Indigenous Media. Cultures, Poetics and Politics*. Duke.

11

Prefigurative Politics: Towards a Cosmic Socialism?

Prefigurative Politics

'Prefigurative politics' as implied in Marx's writings and in much fuller detail by the New Left in our own era is not about a simple 'left' and 'right' dichotomy. It entails a politics which addresses immediate social and political crises while at the same time confronting longer term demands of the social and environmental order. The outcome has been a form of libertarianism linked around a central theme. Raekstand and Gradin refer to prefigurative politics as 'the deliberate experimental implementation of desired future social relations and practices in the here-and-now'.

Prefiguration: The Lucas Aerospace Example

To illustrate what 'prefigurative politics' means, this chapter first turns the famous example in the 1970s when a group of British aerospace workers tried to take over their factory's commitment to save not only their jobs but also to enhance environmental conditions.

In 1974, Lucas Aerospace, like other aerospace companies in Britain and elsewhere, announced a plan to restructure the company as a consequence of increased international competition and technological change. Around half of Lucas Aerospace output supplied military contracts.

Since this depended upon public funding, as did many of the civilian contracts, the Lucas employees argued that state support should be put into making products that were really needed by society at large. This rather than making employees redundant and depending on government handouts.

The result was a plan drawn up by the Lucas Aerospace workers themselves to produce items that were definitely *not* based on wars and capital accumulation to this end. Rather, the priority had to be making products that were urgently needed. This included buses which could be run on roads and rail, heat pumps, solar cell technologies, wind turbines and new fuel cell technologies, and kidney machines. The Lucas aerospace employees believed it was scandalous that people could be dying for the want of a kidney machine when those who could be producing them are facing the prospect of redundancy. So now it was to be employees themselves—and not owners of capital and military authorities—who would decide on and make what was genuinely needed for society as a whole. The project received a sympathetic hearing from the radical Labour government in power at the time, though it did not have the powers to make the necessary legislation.

Forty projects were put forth by the Lucas Aerospace workforce. From them, products were selected which fell into six categories: medical equipment, transport vehicles, improved braking systems, energy conservation, and telechiric machines in which production could be carried out at a safe distance from the object being made. Specific proposals included, in the medical sector and the mass production of kidney dialysis machines, which at that time were being manufactured on one of the Lucas Aerospace sites.

The proposals—again advanced by employees themselves—were rejected out of hand by Lucas Aerospace management. They refused to diversify from aerospace work, even though they knew that it was already in decline. The existence of marginal industrial and medical equipment already being carried out on some of the sites could have been built upon, but the idea of employees themselves deciding what should be made was unsurprisingly unattractive to bosses and shareholders. It looked like workers' control, which of course it was.

The Lucas Aerospace plan made by its employees received worldwide support. Lucas shop stewards attended numerous meetings in the United Kingdom and working visits to Sweden, Germany, Australia and the United States. But, in the end, this project was unsuccessful. While individual Trade Unions and the Labour Government supported the Combine's Plan in principle, there was neither the structures in place, nor the political will, to put pressure on Lucas Aerospace management to negotiate with the Combine. But this is not the end of the story.

The Combine's attempt to make environmentally and socially valuable products was taken up by the Left Wing Greater London Council in 1984. But this too was abandoned when Prime Minister Thatcher closed it down. The whole idea of workers controlling what they should make and using their own labour processes was not exactly popular to a radical right, monetarist, prime minister.

Meanwhile, a radical left-wing Greater London Council drew up plans based on those made by the Lucas Aerospace workers. But the GLC and its plans were terminated by an incoming radical monetarist government headed by Margaret Thatcher. They did not like the idea of employees deciding for themselves what products should be made. It was the world turned upside down and as such represented a threat to the whole social order. So this meant it had to be terminated. It was the end of the Lucas Aerospace project.

But even though the idea of producing socially and environmentally useful products was defeated, it still stands as an objective lesson for a future form of socialism.

Prefiguring Alternatives Now: The Jet Propulsion Laboratory

At the heart of a socialist society would be social relations at the workplace and the full-scale recognition of people's skills. The JPL workers discussed in Chap. 5 have a special position in this context. They prefigure that a new socialist relation between society and the cosmos might be made and might look alike.

There are two linked ways in which the JPL workers should be celebrated and made examples of a future form of socialism. First, their work is undertaken as a means of understanding the cosmos. This, as discussed shortly, is one of the most positive projects imaginable. It is creating new forms of knowledge of the cosmos. But at the same time, new kinds of social relations are formed. These people prefigure new, collectively collaboratively organised ways of making such a project. Furthermore, they are part of a global project, working with other agencies in Canada and Europe.

Second, the JPL workers as a whole have of course contributed to the making of something very special, the James Webb Space Telescope, in collaboration with European and Canadian space agencies. They are making a spacecraft which will genuinely enhance our understanding of the cosmos (Greenhouse, 2012) (Fig. 11.1).

Fig. 11.1 NASA's Webb reveals Cosmic Cliffs, glittering landscape of star birth. Jul 12, 2022

- NASA's James Webb Space Telescope reveals emerging stellar nurseries and individual stars in the Carina Nebula that were previously obscured
- Red arcs in the image trace light from galaxies in the very early universe. The image is said to be the deepest, most detailed infrared view of the universe to date, containing the light from galaxies that has taken many billions of years to reach us.
- Images of 'Cosmic Cliffs' show Webb's cameras' capabilities to peer through cosmic dust, shedding new light on how stars form.
- Objects in the earliest, rapid phases of star formation are difficult to capture, but Webb's extreme sensitivity, spatial resolution and imaging capability can chronicle these elusive events.

Making the James Webb Telescope is an object lesson for everyone who is interested in alternatives other than using the cosmos for purposes of capital accumulation and the enhancement of military power. It shows how an extraordinarily worthwhile scientific project can be made in a *collective* way between nations, a way relatively separated from the demands made by capital on labour forces. The James Webb telescope is an outstanding collaboration of skilled people throughout society.

Earlier I praised the Lucas Aerospace workers who organised and made environmentally and socially valuable products. 'The James Webb', as it is called, is a global and cosmic equivalent, advancing our knowledge of the cosmos at the earliest stage of its evolution, shortly after The Big Bang and the origins of the Universe. It is difficult to imagine a 'product' that is more worthwhile.

It is difficult to imagine a more important project for mankind, and a socialist society would surely give absolute priority to work similar to the one conducted by JPL. And producing the James Webb Telescope makes a central political point. A profitability based on militarisation is a fraud in two ways. First, it means that employees do not sufficiently gain from what they are making. This has long been in the nature of capitalism, one in which profits stemming from employees' work are siphoned off to company executives and shareholders. Second, investors are financing projects which are self-centred, militaristic and extraordinarily dangerous. All this means that profitability based on the militarisation of the cosmos should stay out of the question, this still being the demand by The United Nations.

And again, this is where the James Webb telescope comes into the picture. It, and in due course its successors, enhances our understanding of the cosmos. Furthermore, it is an international project, invented and produced across a range of societies. All these considerations make a project such as James Webb more progressive and life-affirming than grubbing around the cosmos for militaristic and profit-seeking ends.

References and Further Reading

Cooley, M. (2018). *Delinquent Genius. The Strange Affair of Man and His Technology*. Spokesman.
Cooley, M. (2020). *The Search for Alternatives. Liberating Human Imagination*. Spokesman.
Dickens, P. (2009). The Cosmos as Capitalism's Outside. In D. Bell & M. Parker (Eds.), *Space Travel and Culture* (Sociological Review Monograph. Vol. 57, Issue 1).
Greater London Council. (2020). *London Industrial Strategy*. Greater London Authority.
Greenhouse, M. (2012). NASA Technical Reports Server. *The James Webb Space Telescope Mission*. Institute of Aeronautics and Astronautics.
Palmer, J., & Karamjit, S. (2020). *The Search for Alternatives. Liberating Human Imagination. A Mike Cooley Reader*. Spokesman.
Pathak, P. S. (2021). *James Webb Space Telescope. The Future of Space Astronomy*. Amazon.
Raekstad, P., & Gradin, S. S. (2020). *Prefigurative Politics. Building Tomorrow Today*. Polity.
Sharp, G. (2001). *Ecology and the Labour Process: Towards a Prefigurative Sociology of the Labour-Nature Relation*. PhD Thesis, University of Sussex.
Stakem, P.(2021). *The James Webb Space Telescope, a New Era in Astronomy* (Space Series No.37).
Wainwright, H., & Elliott, D. (2018). *The Lucas Plan. A New Trade Unionism in the Making?* Allison and Busby.

Index

A

Aboriginal, 135–137, 139–141, 143, 144
Adult infantile narcissism, 6, 28, 31
Alberti, L., 26
Altruism, 27, 28
Artemis Moon project (NASA), 22
Australia, 137–139, 141, 143–146, 151

B

Bacon, F., 35, 95, 96, 99
Beck, U., 7, 95–98, 100–103
Bezos, J., 17, 18, 20–23, 29, 36
Blue Origin, 22
Body, 5, 7, 68, 73–91, 101, 117–119, 122–125, 136

C

Capital, 3–8, 14–16, 18, 20, 22, 23, 25, 36, 39–53, 58, 71–72, 74, 77, 87, 90, 91, 99, 100, 102–115, 120, 121, 125, 127, 129, 130, 132, 135, 139, 141, 150, 153
  accumulation, 3–8, 18, 21, 22, 41–44, 47–52, 70, 91, 105–115, 129, 136, 150, 153
Capitalism
  Freud, 27
  labour, 15, 41
  rise of, 26
Capitalist, 6, 13, 14, 16, 17, 19, 25–30, 39–42, 49, 50, 58, 59, 97, 102, 106, 108–109, 117, 125, 129, 131, 136–139, 146

Index

Circuits of capital, 39–53
Consumption-levels, 32
Cosmic capitalism, 1–8, 11–23, 27, 29–30, 34, 42, 44, 57, 71, 73–91, 99, 102, 109, 117–132, 135–147
Cosmopolitanism, 26
Cosmos, 1–8, 12, 14–21, 23, 25–36, 39, 41–43, 46, 49–52, 58–61, 67, 68, 71, 73–77, 79–91, 95, 98–99, 101–105, 107–109, 111, 113–115, 117–132, 137, 139, 151–154
Craib, I., 31

D
Dean, K., 32
Desire, 27, 28, 30, 34
Dickens, P., 34, 121
Disappointment, 28, 31, 32
Durkheim, E., 137–139, 146
Dynetics (company), 23

E
Extraction (materials), 16, 43, 91, 107

F
Fantasy, 11, 25–36, 60
Far West, 29
Feudalism, 25
Freud and capitalism, 27
Freud, S., 26–32
Fromm, E., 34

G
Garretson, P., 1–5, 122, 130, 131
Goswami, N., 1–5, 122, 130, 131

H
Heinlein, R.A., 14, 15, 20
Holen, T., 18, 19, 21, 42, 103, 124–126

I
Idealism, 18, 27, 28, 121
Imparja TV, 140–144
Imperialism, 106–109
Indigenous, 136–144, 146
   peoples, 8, 135, 137–147
Individualism, 26, 31

J
James Webb Telescope, 8, 153, 154
Jet Propulsion Laboratory (JPL), 7, 59–68, 70–72, 113, 114, 151–154

L
Libertarianism/space, 19
Libertarian Party, 12, 13
Lockheed Martin, 20, 52, 111, 112

M
Making spacecraft, 7, 58–61, 68, 114

Mars, 4, 7, 8, 29, 36, 59–64, 66–68, 70, 71, 81, 84, 108, 113, 126
Marx, K., 5–8, 13–15, 57, 58, 103, 136, 138, 149
Mean, M., 34
Military, 1, 4–8, 19–21, 36, 39–41, 45, 46, 58–61, 68, 71–75, 80, 87, 91, 100, 103–106, 108–115, 120, 122–124, 131, 150, 153
Mill, J.S., 12
Moon, 3, 4, 8, 14, 16, 21–23, 29, 33–36, 58–60, 63, 75, 77, 81, 87, 88, 91, 99, 100, 107, 108, 118, 119, 122, 123, 126–131
Moon Science, 22

N
Narcissism, 6, 25–36
National Indigenous Television (NITV), 144–146
Northrop Grumman, 20, 23, 52, 111

O
Omnipotence, 28, 30, 32–34
Ormrod, J., 29, 30, 32–35
Outer Space Narcissism, 27–29
Outer Space Treaty (OST), 118–120, 122–131

P
Profit, 4, 6, 18, 34, 40–43, 46, 48–53, 58, 59, 74, 97, 99, 106, 111, 113, 127, 129–131, 135, 143, 153
Pro-space activism, 32–35

R
'Remote' television, 141–144, 146
Renaissance Florence, 26
Renaissance Italy, 26
Rocket(s), 6, 7, 18–20, 22, 57–60, 63, 65, 68, 71–72, 75, 85, 86, 98, 100–102, 108, 112, 115
Rotation of capital, 7, 48, 50, 51

S
Satellisation of society, 7, 39
Satellite/satellites, 4, 6–8, 19, 21, 22, 39–52, 57, 60, 86, 89, 91, 98–100, 102, 105–115, 119, 120, 130, 135, 140, 142
  military, 46, 60, 91, 112
  television, 7, 8, 135–147
Satnavs, 7, 41
Science fiction, 14, 33, 64
Self-absorption, 29, 30, 32
Self expression, 27, 28
Self-identity, 25
Self-love, 27
Sennett, R., 30
Shammas, V., 18, 19, 21, 42, 103, 124–126
Socialism, 8, 149–154
Sorensen, C., 20, 110–113

Spacecraft, 4, 6, 7, 18, 20, 21,
    41, 42, 57–65, 68–70,
    78, 85, 89, 90, 98, 99,
    101, 108, 112, 114,
    119, 152
    production, 40, 57, 58,
        68, 71, 114
Space junk, 7, 21, 46, 98, 99, 101,
    102, 111
Space law, 8, 117–132
Space race, 1, 76
SpaceX, 19, 22
'Starlink' project, 41

T
Television (TV), 6–8, 19, 49, 50, 59,
    61, 70, 71, 135–147
    programming, 135, 139, 141

Turner, F.J., 11

U
United Launch Alliance (ULA),
    17, 60–61, 68–70,
    114, 115
United Nations (UN), 8, 101,
    117–125, 130–132, 153

W
Wallerstein, I, 25
West Frontier' (USA), 11
Wilsdon, J., 34

Z
Zubrin, R., 11

Printed by Printforce, the Netherlands